1994

Current Topics in Membranes, Volume 41

Cell Biology and Membrane Transport Processes

Current Topics in Membranes, Volume 41

Yale Series Editors

Joseph F. Hoffman and Gerhard Giebisch
Department of Cellular and Molecular Physiology
Yale University School of Medicine
New Haven, Connecticut

Murdoch Ritchie
Department of Pharmacology
Yale University School of Medicine
New Haven, Connecticut

Series Editors

Arnost Kleinzeller
Department of Physiology
University of Pennsylvania
School of Medicine
Philadelphia, Pennsylvania

Douglas M. Fambrough
Department of Biology
The Johns Hopkins University
Baltimore, Maryland

Current Topics in Membranes, Volume 41

Cell Biology and Membrane Transport Processes

Guest Editor

Michael Caplan

Department of Physiology
Yale University School of Medicine
New Haven, Connecticut

Volume 41 is part of the series from the
Yale Department of Cellular and Molecular Physiology.

ACADEMIC PRESS
A Division of Harcourt Brace & Company
San Diego   New York   Boston   London   Sydney   Tokyo   Toronto

Copyright © 1994 by ACADEMIC PRESS, INC.
All Rights Reserved.
No part of this publication may be reproduced or transmitted in any form or by any means, electronic or mechanical, including photocopy, recording, or any information storage and retrieval system, without permission in writing from the publisher.

Academic Press, Inc.
525 B Street, Suite 1900, San Diego, California 92101-4495

United Kingdom Edition published by
Academic Press Limited
24–28 Oval Road, London NW1 7DX

International Standard Serial Number: 0070-2161

International Standard Book Number: 0-12-153341-7

PRINTED IN THE UNITED STATES OF AMERICA
94 95 96 97 98 99   EB   9 8 7 6 5 4 3 2 1

Contents

Part I The Multidrug Resistance Family of Transporters

CHAPTER 1 The Multidrug Transporter: Mechanistic Considerations

Michael M. Gottesman, Stephen Currier, Edward Bruggemann, Isabelle Lelong, Wilfred Stein, and Ira Pastan

CHAPTER 2 A Novel Mechanism for Transmembrane Translocation of Peptides: The *Saccharomyces cerevisiae STE6* Transporter and Export of the Mating Pheromone **a**-Factor

Karl Kuchler, Elana E. Swartzman, and Jeremy Thorner

Part II Structure–Function Relationships in Ion Pumps

CHAPTER 3 Structural Requirements for Subunit Assembly of
the Na,K-ATPase

*Douglas M. Fambrough, M. Victor Lemas, Kunio Takeyasu,
Karen J. Renaud, and Elizabeth M. Inman*

CHAPTER 4 Structure–Function Relationship of Na,K-ATPase:
The Digitalis Receptor

*Cecilia Canessa, Frédéric Jaisser, Jean-Daniel Horisberger,
and Bernard C. Rossier*

CHAPTER 8 Synthesis and Sorting of Ion Pumps in Polarized Cells

Cara J. Gottardi, Grazia Pietrini, Monica J. Shiel, and Michael J. Caplan

Contributors

Numbers in parentheses indicate the pages on which the authors' contributions begin.

Qais Al-Awqati (109), Department of Medicine, College of Physicians and Surgeons, Columbia University, New York, New York 10032

Edward Bruggemann (3), Laboratory of Molecular Biology, National Cancer Institute, National Institutes of Health, Bethesda, Maryland 20892

Cecilia Canessa (71), Institut de Pharmacologie et de Toxicologie de l'Université, CH1005 Lausanne, Switzerland

Michael J. Caplan (143), Department of Cellular and Molecular Physiology, Yale University School of Medicine, New Haven, Connecticut 06510

Stephen Currier (3), Laboratory of Cell Biology, National Cancer Institute, National Institutes of Health, Bethesda, Maryland 20892

Douglas M. Fambrough (45), Department of Biology, The Johns Hopkins University, Baltimore, Maryland 21218

Cara J. Gottardi (143), Department of Cellular and Molecular Physiology, Yale University School of Medicine, New Haven, Connecticut 06510

Michael M. Gottesman (3), Laboratory of Cell Biology, National Cancer Institute, National Institutes of Health, Bethesda, Maryland 20892

Peter M. Haney (89), Department of Pediatrics, Washington University School of Medicine, St. Louis, Missouri 63110

Jean-Daniel Horisberger (71), Institut de Pharmacologie et de Toxicologie de l'Université, CH1005 Lausanne, Switzerland

Elizabeth M. Inman (45), Department of Biology, The Johns Hopkins University, Baltimore, Maryland 21218

Frédéric Jaisser (71), Institut de Pharmacologie et de Toxicologie de l'Université, CH1005 Lausanne, Switzerland

Karl Kuchler (19), Department of Molecular and Cell Biology, Division of Biochemistry and Molecular Biology, University of California, Berkeley, California 94720

Isabelle Lelong (3), Laboratory of Cell Biology, National Cancer Institute, National Institutes of Health, Bethesda, Maryland 20892

M. Victor Lemas (45), Department of Biology, The Johns Hopkins University, Baltimore, Maryland 21218

Mike Mueckler (89), Department of Cell Biology and Physiology, Washington University School of Medicine, St. Louis, Missouri 63110

W. James Nelson (123), Department of Molecular and Cellular Physiology, Stanford University School of Medicine, Stanford, California 94305

Ira Pastan (3), Laboratory of Molecular Biology, National Cancer Institute, National Institutes of Health, Bethesda, Maryland 20892

Grazia Pietrini (143), CNR Department of Pharmacology, Chemotherapy and Medical Toxicology, University of Milan, 20129 Milan, Italy

Karen J. Renaud (45), Department of Biology, The Johns Hopkins University, Baltimore, Maryland 21218

Bernard C. Rossier (71), Institut de Pharmacologie et de Toxicologie de l'Université, CH1005 Lausanne, Switzerland

Monica J. Shiel (143), Department of Cellular and Molecular Physiology, Yale University School of Medicine, New Haven, Connecticut 06510

Wilfred Stein (3), Laboratory of Molecular Biology, National Cancer Institute, National Institutes of Health, Bethesda, Maryland 20892

Elana E. Swartzman (19), Department of Molecular and Cell Biology, Division of Biochemistry and Molecular Biology, University of California, Berkeley, California 94720

Kunio Takeyasu (45), Department of Biology, The Johns Hopkins University, Baltimore, Maryland 21218

Jiro Takito (109), Department of Medicine, College of Physicians and Surgeons, Columbia University, New York, New York 10032

Jeremy Thorner (19), Department of Molecular and Cell Biology, Division of Biochemistry and Molecular Biology, University of California, Berkeley, California 94720

Janet van Adelsberg (109), Department of Medicine, College of Physicians and Surgeons, Columbia University, New York, New York 10032

Preface

Membrane transport processes play a critical role in determining the composition of the intracellular and extracellular environments. Recent advances in cellular and molecular physiology have greatly expanded our knowledge of the structure and regulation of the proteins that mediate these transport functions. It has become apparent that a complete understanding of membrane transport processes must include insight into both the structure and the function of the transporters themselves as well as into the cellular mechanisms that orchestrate the transporters' biosynthesis and subcellular distribution. The ability of epithelial cells, for example, to carry out vectorial resorption or secretion is dependent upon the strict segregation of specific membrane transport systems to one or the other of their two plasmalemmal domains. Similarly, a muscle cell's ability to respond to insulin depends upon the appropriate sorting and mobilization of a class of glucose transport proteins. Furthermore, it has become clear that a number of processes which have traditionally been thought of as "cell biologic" (i.e., protein secretion) may be best understood in the context of new physiologic information on membrane transport systems. The purpose of this volume is to explore the interface between membrane transport and cell biologic processes and to examine the implications of their emerging inter-relationships.

The volume's first section is devoted to a new class of membrane transport proteins that share structural homology and appear to perform a diverse array of tasks. The multidrug transporter (MDR), cystic fibrosis transmembrane conductance regulator (CFTR), and the yeast protein STE 6 appear to exploit common structural motifs in the transport of xenobiotics, chloride, and small peptides, respectively. The first two chapters discuss the cell biologic and physiologic significance of these transporters, as well as the functional features that unite and distinguish them.

The second section focusses on a family of ion pumps involved in ATP-dependent cation transport. Recently, a great deal has been learned through the application of molecular techniques to identify structural features of these molecules important for determining their cell biologic and physiologic properties. These chapters describe the identification of structural motifs responsible for ion selectivity, inhibitor sensitivity, and sub-unit assembly.

The third section is devoted to transport protein sorting and recycling. Ion transport proteins can only subserve their appropriate physiologic functions if they are directed to and maintained within the appropriate subcellular domains. The last four chapters investigate the signals and cellular mechanisms that contribute to the localization of the insulin re-sponsive glucose transporter, the renal proton pump, and ion pumps of the Na,K-ATPase family.

I acknowledge the generous financial support of Boehringer Ingelheim Pharmaceuticals, Inc.; Miles, Inc.; Sandoz Research Institute; Burroughs Wellcome Company; Lilly Research Laboratories; Packard Instrument Company; The Upjohn Company; and ICI Pharmaceuticals Group, which made the conference from which this volume is derived possible. I also thank Drs. J. F. Hoffman and G. H. Giebisch for inviting me to organize the conference and edit the resulting volume. Finally, I thank Dr. Lorraine Lica and Ellen Caprio for their patience and assistance in organizing the text.

MICHAEL J. CAPLAN

Previous Volumes in Series

Current Topics in Membranes and Transport

* Part of the series from the Yale Department of Cellular and Molecular Physiology

PART I

.

The Multidrug Resistance Family of Transporters

CHAPTER 1

The Multidrug Transporter: Mechanistic Considerations

Michael M. Gottesman, Stephen Currier, Edward Bruggemann,*
Isabelle Lelong, Wilfred Stein,* and Ira Pastan*
Laboratory of Cell Biology, and *Laboratory of Molecular Biology, National Cancer Institute, National Institutes of Health, Bethesda, Maryland 20892

I. INTRODUCTION: THE PROBLEM OF MULTIDRUG RESISTANCE IN CANCER

For some disseminated human cancers, such as leukemias, lymphomas, testicular cancer, choriocarcinoma, and childhood sarcomas, treatment with combinations of cytotoxic chemotherapeutic drugs eliminates disease in a substantial number of patients. Other cancers, such as common solid tumors of the lung, colon, liver, pancreas, and kidney, respond poorly to chemotherapy, and many cancers that do respond initially to chemotherapy later relapse and become unresponsive. Because it is possible to cure some cancers with combination chemotherapy, there is reason to hope that if the mechanisms of intrinsic and acquired chemotherapy were understood, strategies to circumvent drug resistance could be developed and used to improve the treatment of drug-resistant human cancers.

In recent years, one form of simultaneous resistance to multiple chemotherapeutic drugs has been shown to be due to expression of an energy-dependent drug efflux pump in the plasma membranes of resistant cells (reviewed in Gottesman and Pastan, 1988, 1993; Endicott and Ling, 1989). By this mechanism, a number of different anticancer drugs are pumped out of resistant cells. These drugs, enumerated in Table I, are structurally quite distinct but share the following properties: (1) they are relatively hydrophobic, and thus have partition coefficients which favor lipid over aqueous solubility; (2) most are positively charged at physiological pH; and (3) virtually all are natural products of plants or microorganisms, or are semi-synthetic derivatives of such compounds. Not all anticancer drugs are affected by this pump system. Many effective drugs, including nucleoside analogs, alkylating agents, and *cis*-platinum, are not involved in this multidrug resistance phenotype.

II. GENERAL STRUCTURAL FEATURES OF THE MULTIDRUG TRANSPORTER

The cloning and sequencing of a cDNA for the multidrug transporter from the mouse (Gros *et al.*, 1986) and human (Chen *et al.*, 1986) made it possible to generate a model for its primary structure. The human

TABLE I

Cytotoxic Drugs to Which MDR Cells Are Resistant

Anticancer drugs	Other agents
Vinca alkaloids (vinblastine, vincristine)	Colchicine
	Puromycin
Anthracyclines (daunorubicin, doxorubicin, mitroxantrone)	Podophyllotoxin
	Ethidium bromide
Epipodophyllotoxins (etoposide, teniposide)	Emetine
Mitomycin C	Gramicidin D
Actinomycin D	Valinomycin
Taxol	
Topotecan	
Mithramycin	

multidrug transporter, also known as P-glycoprotein, has a molecular mass of 170,000 daltons, is glycosylated, and consists of 1280 amino acids encoded by the *MDR*1 gene. The sequence analysis predicts 12 membrane-spanning domains, with 6 in the amino-terminal and 6 in the carboxy-terminal halves of the molecule (see Fig. 1). Although the localization of the transmembrane domains has not been totally confirmed by independent biochemical, genetic, or histochemical analysis, most of the available data support the general features of this model (Gottesman *et al.*, 1991). The glycosylation appears limited to the first extracytoplasmic loop, with three potential sites indicated by the primary sequence, and no other potential sites of glycosylation have been found to be glycosylated in forms of the molecule found on the surface of cells (Bruggemann *et al.*, 1989). The molecule has predicted symmetrical structures around an amino acid region which appears to connect, or link, the two halves of the molecule, and there is 43% sequence identity between the two halves.

Although there is very little sequence identity in the transmembrane regions between the various forms of P-glycoprotein and other transmembrane proteins, comparison of the sequence within the two large cytoplasmic loops with sequences in the data base shows considerable sequence identities in the region circled in Fig. 1 (top). This conserved cytoplasmic region appears to be involved in the binding of ATP and its utilization as a source of energy in the transport process. A large number of proteins have now been identified which share this ATP-binding cassette (ABC), most of which are membrane transport proteins, or subunits of multiprotein transport complexes (Higgins *et al.*, 1986). This ABC family includes nutrient transporters in bacteria, pigment transporters in *Drosophila,* and polypeptide transporters in bacteria, yeast, and mammals. There are also several new family members whose precise function is not yet known, including the protein defective in cystic fibrosis (CFTR) (Riordan *et al.*, 1989) and the major peroxisomal membrane protein in liver (Kamijo *et al.*, 1990).

Searches for proteins with sequence similarities to P-glycoprotein outside of the ATP-binding regions have identified many such proteins. In the human, there is a second *MDR* gene known as *MDR*2 which encodes a protein over 80% identical in predicted amino acid sequence to *MDR*1. Only the *MDR*1 gene encodes a functional multidrug transporter. The *MDR*2 gene (also known as *mdr*3) is expressed in liver (Van der Bliek *et al.*, 1988) and is presumed to encode another transporter of homologous function, but its substrates are yet to be determined. In rodents there are two functional *mdr*1-type genes as well as an *mdr*2 gene. *Drosophila, Falciparum* malaria, *Caenorhabditis elegans,* and *Leishmania* have one or more genes closely related to *mdr* (Schinkel and Borst, 1991). In the

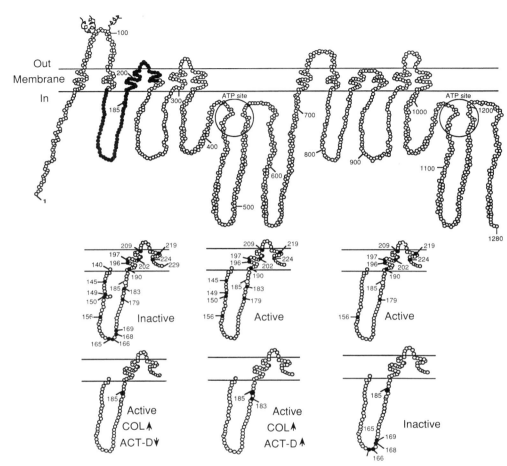

FIGURE 1 Predicted structure of P-glycoprotein. (Top) A schematic drawing based on the hydrophobicity profiles and potential ATP-binding sites deduced from the primary sequence (Chen *et al.*, 1986). The circles indicate the putative ATP-binding sites, the wiggly lines show potential glycoprotein sites, and the Gly to Val mutation at position 185 which alters substrate specificity is shown. (After Gottesman and Pastan, 1988.) The filled circles show the region of chimerism in which MDR2 sequences were substituted for *MDR*1 sequences. (Bottom) The first intracytoplasmic loop mutations in *MDR*1/*MDR*2 chimeras in this region. The filled circles represent mutations either introduced in the chimera or otherwise affecting this region. (After Currier *et al.*, 1992). The effect on substrate specificity of these mutations is shown for each construct as determined by measuring patterns of drug resistance in NIH3T3 cells transfected with pHaMDR1/A vectors (Pastan *et al.*, 1988) carrying these mutations (Currier *et al.*, 1992).

case of the malarial parasite, there is some published evidence suggesting that this *mdr* gene is responsible for chloroquine resistance (Foote *et al.*, 1990), but function in the other organisms has not yet been determined.

III. EVIDENCE INDICATING THAT P-GLYCOPROTEIN IS NOT A CLASSICAL TRANSMEMBRANE TRANSPORTER SUGGESTS A NEW MODEL FOR DRUG EFFLUX

Many lines of evidence have converged to suggest that the multidrug transporter does not function as a simple transmembrane transport system which pumps drugs out of the cytoplasm of resistant cells into the extracellular space. First, as already noted, most of the pump's substrates are hydrophobic and partition into apolar solvents in preference to aqueous phases (Zamora *et al.*, 1988), suggesting that substrates for P-glycoprotein may be concentrated in the plasma membrane in cells which do not have the pump. Use of confocal epifluorescence microscopy has confirmed that, in drug-sensitive cells, several P-glycoprotein substrates, including the anticancer anthracyclines and the mitochondrial laser dye rhodamine 123, are predominantly localized in the plasma membrane and intracellular membranous structures in addition to their normal cytoplasmic targets (Weaver *et al.*, 1991).

Second, the wide variety of drugs handled by the transporter, including amphipathic hydrophobic natural products and hydrophobic peptides (gramicidin D, valinomycin, and cyclosporine A), suggests that a simple model of substrate recognition and transport is not likely to be correct. The critical identifying features of substrates seem to be physical (hydrophobicity) rather than chemical. Many of these cytotoxic drugs have intracellular targets, such as tubulin or DNA, to which they bind tightly. At steady state, for a pump to be able to remove cytoplasmic or nuclear drug from a cell treated with a cytotoxic agent, it would have to compete with the cytotoxic target (such as tubulin or DNA) to which the drug binds. It is difficult to imagine a single pump having greater affinity for all of these substrates than their respective cytotoxic targets.

Third, detailed kinetic analyses of drug accumulation and efflux by cells expressing the multidrug transporter consistently suggest that there is a decrease in drug influx as well as an increase in efflux (reviewed in Stein *et al.*, 1993). One interpretation of these results is that decreased influx results from genetic changes unrelated to expression of the *mdr* gene in some of these cell lines. However, the widespread occurrence of these decreases in influx whenever P-glycoprotein is found suggests that the multidrug transporter affects both influx and efflux (Stein *et al.*, 1993).

Thus, the kinetic data are not consistent with the idea that P-glycoprotein is a simple transmembrane transporter.

Finally, a recent report associates the development of chloride channels regulated by osmotic swelling with expression of P-glycoprotein (Valverde *et al.*, 1992), and, similarly, another ABC family member, CFTR, appears to have chloride-channel activity (Kartner *et al.*, 1991; Anderson *et al.*, 1991). If the chloride-channel activity is related to P-glycoprotein-dependent drug transport, than current models of P-glycoprotein action will need to be revised.

We have proposed the general model shown in Fig. 2 to account for many of these paradoxical observations about the multidrug transporter (Raviv *et al.*, 1990; Gottesman *et al.*, 1991). The principle feature of this model is that drugs are removed by the transporter directly from the plasma membrane. This enables the pump to intercept drugs before they reach the cytoplasm. Although most of the substrates for P-glycoprotein are amphipathic, with a positive charge at physiological pH, there is reasonable evidence based on the pH dependence of the uptake of drugs in drug-sensitive cells, and on the relationship between permeability and molecular weight (Stein *et al.*, 1993), to suggest that the neutral form of the drug diffuses through the membrane by simple diffusion (Dalmark and

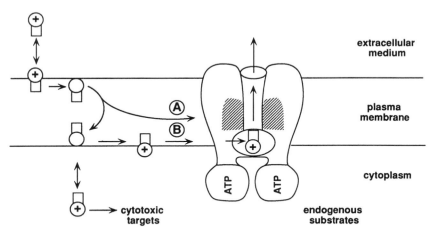

FIGURE 2   Model indicating that the multidrug transporter probably detects hydrophobic drugs within the lipid bilayer. Two alternate routes of transport are indicated: (A) an uncharged species which is trapped by a proton within the transporter; (B) a charged species which is moved through the membrane by the transporter. Other mechanistic features illustrated here include the contribution of drug-labeling sites from two halves of the molecule to form a single site (cross-hatched area) and the ATP utilization site(s). (After Gottesman *et al.*, 1991.)

Storm, 1981). The form of the drug visualized in the plasma membrane by confocal epifluorescence microscopy (Weaver *et al.*, 1991) is either the neutral form or a charged form embedded in either the inner or outer leaflet of the bilayer. Either of these forms could be substrates for the transporter.

What is the evidence that P-glycoprotein can remove drugs directly from the lipid bilayer? Epiflourescence excitation data (Kessel, 1989) and confocal epifluorescence microscopy (Weaver *et al.*, 1991) suggest that the substrate rhodamine 123 is in an apolar environment (membranes) in drug-sensitive cells and in an aqueous environment (extracellular space) in drug-resistant cells. Energy transfer studies using the photoaffinity label iodonaphthalene azide and the chromophoric drug daunorubicin demonstrate that P-glycoprotein effectively extracts daunorubicin from its membrane locations in drug-sensitive cells (Raviv *et al.*, 1990).

The general idea is that the multidrug transporter acts as a "hydrophobic vacuum cleaner" to remove drugs from the membrane with energy provided by ATP. How is the energy of ATP linked to the actual translocation process? This is clearly a major question for future study, but several intriguing hypotheses can be considered: (1) The transporter acts as a trap for a neutral drug by giving it a positive charge. This could happen if the transporter is a protontropic engine which pumps protons into a confined aqueous space defined by P-glycoprotein within the plasma membrane. Several published reports suggest that activity of the transporter is associated with extrusion of a proton (Thiebaut *et al.*, 1990; Keizer and Joenje, 1989; Boscoboinik *et al.*, 1990). Once trapped within the transporter in the membrane, the drug can no longer cross the membrane but can be released from the transporter into the extracellular space (Stein *et al.*, 1993). This mechanism could work only for drugs which acquire a positive charge at neutral or acid pH. (2) The multidrug transporter is essentially a "flippase" which takes positively charged drug from the inner leaflet and flips it to the outer leaflet, or flips it into the extracellular space (Higgins and Gottesman, 1992). For this model to be correct there would need to be substantial amounts of charged drug embedded in the plasma membrane, and it is unclear if this occurs. (3) The pump is essentially a constantly moving waterwheel, or an escalator, which expels all membrane constituents of approximately the correct molecular weight and shape, with little chemical specificity. There are currently no data which distinguish clearly among these hypotheses, but the recent suggestion of chloride-channel activity associated with the multidrug transporter is consistent with the idea that a net positive charge accompanies drugs out of the cell and that this may require an anion channel so that the transport process remains electrically neutral.

The following sections briefly consider the evidence for other features of the model shown in Fig. 2, including the location of the sites labeled by photoactivatable drugs; the evidence that these sites come from both halves of the molecule but are close to each other in the three-dimensional structure, sites involved in substrate specificity; and the requirement for ATP in the transport process.

IV. IDENTIFICATION OF PHOTOLABELED SITES IN P-GLYCOPROTEIN

Several radioactive, photoactivatable substrates for the multidrug transporter have been utilized to show that the transporter binds drugs directly and to begin to identify the labeled sites in the transporter. Photoactive substrates include an iodinated derivative of vinblastine, [^{125}I]NASV (Cornwell *et al.*, 1986,1991; Safa *et al.*, 1986) a tritiated derivative of a calcium-channel blocker, [^{3}H]azidopine (Safa *et al.*, 1987), and radiolabeled derivatives of verapamil (Safa, 1988) and colchicine (Safa *et al.*, 1989). We, and others, have used [^{3}H]azidopine to identify photolabeled sites on the multidrug transporter (Bruggemann *et al.*, 1989; Yoshimura *et al.*, 1989; Greenberger *et al.*, 1990,1991). Some advantages of [^{3}H]azidopine are its commercial availability, the stability of its tritium label, and the finding that its labeling can be inhibited by other substrates for the multidrug transporter, including vinblastine. This latter observation suggests that the sites with which azidopine and vinblastine interact are the same, or very similar, but does not rule out noncompetitive inhibition by vinblastine of azidopine binding (Tamai and Safa, 1991).

Because it has not been possible to obtain purified P-glycoprotein in amounts sufficient to allow direct sequencing of labeled fragments, we have identified [^{3}H]azidopine-labeled fragments by using antibodies specific for different regions of P-glycoprotein (Bruggemann *et al.*, 1989). These antibodies are monoclonal antibodies, antipeptide antibodies (Richert *et al.*, 1988), or obtained by using polypeptide fragments of P-glycoprotein synthesized in *Escherichia coli* (Tanaka *et al.*, 1990; Bruggemann *et al.*, 1991). The strategy has been to photolabel P-glycoprotein with [^{3}H]azidopine in intact cells (so as to preserve the *in vivo* structure of the transporter), immunoprecipitate this labeled protein, digest the immunopurified protein with proteases, or cyanogen bromide, and identify the labeled fragments by immunoprecipitation with antisera specific to known regions of the protein (Bruggemann *et al.*, 1989).

The results of these experiments are summarized in Fig. 3. Both the amino-terminal and carboxy-terminal halves of P-glycoprotein are labeled. Partial digestions with trypsin or V8 protease indicate that in each half of

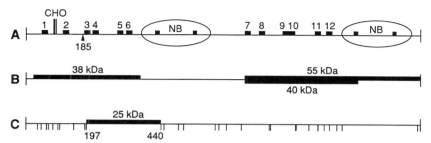

FIGURE 3 Schematic diagram showing [³H]azidopine photolabeled sites in P-glycoprotein determined by partial digestion with trypsin (B) or complete digestion with cyanogen bromide (C). Line A shows a linear diagram of P-glycoprotein in which the transmembrane regions are boxed and numbered, the nucleotide-binding folds (NB) are circled, glycosylations (CHO) sites are shown, and the Gly → Val mutation at residue 185 is indicated. (After Bruggemann *et al.*, 1992.)

the protein the regions including the transmembrane domains and their associated cytoplasmic and extracellular loops, but not the ATP-binding sites, are labeled by [³H]azidopine. In order to quantitate the relative amounts of label in each half accurately, it was necessary to perform limit digestions with cyanogen bromide (Bruggemann *et al.*, 1992). This digestion gives one large amino-terminal fragment (see Fig. 3) which includes approximately 50% of the label in the whole molecule, indicating that the other 50% is in the carboxy-terminal half (whose cyanogen bromide fragments are much smaller).

These quantitative data suggest that the labeled sites in both halves of the molecule are equivalent, but do not distinguish between two identical distinct sites, and one site formed by contributions from each half of the molecule. To help resolve this question, labeling was inhibited with vinblastine over a wide range of concentrations and the distribution of the label in the carboxy-terminal and amino-terminal halves of the molecule was determined. At all concentrations of vinblastine tested, the equal distribution of label in each half of the molecule was observed. These data are most consistent with a single labeling site for [³H]azidopine formed by both halves of the molecule, as shown in Fig. 2.

V. SUBSTRATE RECOGNITION REGIONS AND LABELING SITES ARE DIFFERENT

Although the multidrug transporter is relatively promiscuous with respect to substrate choice (see Table I), its choice of hydrophobic substrates is not entirely random. This point is made apparent by consideration of a mutation in the transporter which arose during sequential selection of human KB cells with colchicine, a relatively poor substrate (Akiyama *et al.*, 1985). At one step in selection, a clone of cells appeared which had increased resistance to colchicine, but was actually less resistant to vinblastine than cells from a previous step. The demonstration that this cell line expressed a P-glycoprotein with a Gly to Val mutation at position 185 (shown in Figs. 1 and 3) (Choi *et al.*, 1988), and that introduction of this mutation reproducibly changed the substrate specificity of the transporter (Choi *et al.*, 1988; Kioka *et al.*, 1989), provided strong evidence that structural changes in the transporter could affect substrate specificity. Furthermore, this mutation lies in the first intracytoplasmic loop of P-glycoprotein outside of the region photolabeled by [³H]azidopine (see Fig. 3), indicating that labeling sites and substrate specificity sites on the molecule may be different.

A more recent analysis of chimeric P-glycoproteins helps to define the function of the first intracytoplasmic loop in determining pump specificity. This region was chosen for further study because of the Gly to Val mutation at residue 185, because deletions of this loop produce a nonfunctional transporter (Currier *et al.*, 1989), and because this region appears to be highly conserved among functional human, mouse, hamster, and rat P-glycoproteins. A segment of P-glycoprotein including the first intracytoplasmic loop, all of transmembrane domain 3, the second extracellular loop, and most of transmembrane domain 4 was excised and replaced with the homologous region from *MDR*2 (Fig. 1). Of 89 amino acids replaced in this region, there are 17 differences between *MDR*1 and *MDR*2.

The resulting chimeric *MDR*1/*MDR*2 transporter is not functional when tested in transfection assays, confirming the essential nature of this region of P-glycoprotein. However, replacement of as few as 4 of the 17 amino acids in the first intracytoplasmic loop region of the chimera with *MDR*1 residues results in a functional transporter (Fig. 1, bottom). These results point out the similarity of the *MDR*1- and *MDR*2-encoded proteins and argue that the *MDR*2-encoded molecule may also be a functional transporter, albeit with different substrate specificity than that of the P-glycoprotein. Furthermore, the essential difference between the two molecules in the region of chimerism lies within the first intracytoplasmic loop, which previous evidence had suggested was involved in substrate specificity.

One of the least conserved changes in the *MDR*1/*MDR*2 chimera is the replacement of an Asn for a Ser at position 183. When a Ser is introduced at position 183 in a molecule which already has a Gly to Val mutation at position 185, substrate specificity is changed once again (Currier *et al.*,

1992). Although colchicine remains a favored substrate, vinblastine and actinomycin D transport also increase (see Fig. 1, bottom). These results point out the involvement of the first intracytoplasmic loop in substrate specificity and support a model in which different residues in this region differentially affect interactions with different substrates. One possible model which accounts for these results is that the mutations in the first intracytoplasmic loop affect a "filter" in the transporter through which substrates must pass before they can enter. An alternate model, favored by data suggesting that the Gly to Val mutation actually increases labeling of the transporter by a derivative of vinblastine (Safa *et al.*, 1990), whose transport is decreased, suggests that the first intracytoplasmic loop is involved in release of substrate *after* it has entered the transporter.

VI. ATP IS THE PREFERRED ENERGY SOURCE FOR *IN VITRO* TRANSPORT

The sequence analysis of P-glycoprotein suggested that it contained two ATP-binding sites, one in each half of the molecule (Gros *et al.*, 1986; Chen *et al.*, 1986) (Fig. 1). Both of these sites appear to be essential for function, as indicated by studies in which amino acid substitutions in either of these sites inactivate the transporter (Azzaria *et al.*, 1989). Preliminary data also suggest that in most cases these sites are not interchangeable, since substitution of either site for the other one usually produces an inactive transporter (S. Currier, U. Germann, S. Kane, I. Pastan, and M. M. Gottesman, unpublished observations). Why does the transporter contain two ATP-binding regions, and is ATP really the preferred energy source for transport?

Currently, the first of these questions is unanswered, but several hypotheses may be entertained. One of the sites may be regulatory, and the second could be catalytic. Both sites might be needed because the first gets the substrate into the transporter, and the second gets it out (I. Roninson, personal communication). The sites may interact with each other to increase the efficiency of transport, with one site functioning as an ATPase, while the second binds ADP released from the first site after hydrolysis. This idea is supported by analysis of other ABC family members, in which the ATP-binding subunits are fused or closely associated with each other (Higgins *et al.*, 1986). In addition to the ABC family of transporters, many other ATPases are now known to have two potential ATP-binding sites, but in no case has their function been clearly defined.

Are these actually ATP-binding and utilization sites? P-glycoprotein has been shown to bind 8-azido-[^{32}P]ATP (Cornwell *et al.*, 1987,1991).

Monoclonal antibodies which recognize eptitopes close to the putative ATP-binding sites block binding of ATP (Georges *et al.*, 1991). To study whether ATP provides energy for the transport process, it was necessary to develop an *in vitro* system for analysis of transport. Using inside-out plasma membrane vesicles derived from multidrug-resistant cells, it is possible to demonstrate ATP-dependent, osmotically sensitive transport of [^{3}H]vinblastine (Horio *et al.*, 1988). In a recent analysis, we asked whether other nucleotides could support vinblastine transport into these vesicles. The results indicate that ATP is the preferred energy source, but GTP works to some extent; ITP and ADP may show some marginal activity; and CTP, UTP, AMP, NAD, NADP, NADH, and NADPH do not support transport (Lelong *et al.*, 1992).

VII. CONCLUSIONS

The investigation of how human cancers elude chemotherapy has uncovered the existence of an energy-dependent multidrug transporter. This transporter belongs to a superfamily of proteins, many of which are other transporters which use the energy of ATP to move drugs, peptides, pigments, nutrients, and other metabolites across cell membranes. Analysis of the mechanism of action of the multidrug transporter reveals some unusual mechanistic features, including accumulating evidence that it recognizes and extrudes drugs directly from the plasma membrane. Continued analysis of the multidrug transporter should yield important information for the rational treatment of drug-resistant cancers and may provide basic knowledge to improve understanding of the entire superfamily to which it belongs.

Acknowledgments

We gratefully acknowledge the help of Joyce Sharrar in the design of the figures and the help of Dwayne Eutsey in typing the manuscript and constructing the figures.

References

Akiyama, S.-i., Fojo, A., Hanover, J. A., Pastan, I., and Gottesman, M. M. (1985). Isolation and genetic characterization of human KB cell lines resistant to multiple drugs. *Somatic Cell Mol. Genet.* **11,** 117–126.

Anderson, M. P., Gregory, R. J., Thompson, S., Sonza, D. W., Paul, S., Mulligan, R. C., Smith, A. E., and Welsh, M. J. (1991). Demonstration that CFTR is a chloride channel by alteration of its anion selectivity. *Science* **253,** 202–205.

Azzaria, M., Schurr, E., and Gros, P. (1989). Discrete mutations introduced in the predicted nucleotide-binding sites of the *mdr*1 gene abolish its ability to confer multidrug resistance. *Mol. Cell. Biol.* **9,** 5289–5297.

Boscoboinik, D., Gupta, R. S., and Epand, R. M. (1990). Investigation of the relationship between altered pH and multidrug resistance in mammalian cells. *Br. J. Cancer* **61**, 568–572.

Bruggemann, E. P., Germann, U. A., Gottesman, M. M., and Pastan, I. (1989). Two different regions of P-glycoprotein are photoaffinity labeled by azidopine. *J. Biol. Chem.* **264**, 15483–15488.

Bruggemann, E. P., Chaudhary, V., Gottesman, M. M., and Pastan, I. (1991). *Pseudomonas* exotoxin fusion proteins are potent immunogens for raising antibodies against P-glycoprotein. *Biotechniques* **10**, 202–209.

Bruggemann, E. P., Currier, S. J., Gottesman, M. M., and Pastan, I. (1992). Characterization of the azidopine and vinblastine binding site of P-glycoprotein. *J. Biol. Chem.* **267**, 21020–21026.

Chen, C.-j., Chin, J. E., Ueda, K., Clark, D., Pastan, I., Gottesman, M. M., and Roninson, I. B. (1986). Internal duplication and homology with bacterial transport proteins in the *mdr*1 (P-glycoprotein) gene from multidrug-resistant human cells. *Cell* **47**, 381–389.

Choi, K., Chen, C., Kriegler, M., and Roninson, I. B. (1988). An altered pattern of crossresistance in multidrug-resistant human cells results from spontaneous mutations in the *mdr*1 (P-glycoprotein) gene. *Cell* **53**, 519–529.

Cornwell, M. M., Safa, A. R., Felsted, R. L., Gottesman, M. M., and Pastan, I. (1986). Membrane vesicles from multidrug-resistant human cancer cells contain a specific 150- to 170-kDa protein detected by photoaffinity labeling. *Proc. Natl. Acad. Sci. U.S.A.* **83**, 3847–3850.

Cornwell, M. M., Tsuruo, T., Gottesman, M. M., and Pastan, I. (1987). ATP-binding properties of P-glycoprotein from multidrug resistant KB cells. *FASEB J.* **1**, 51–54.

Cornwell, M. M., Pastan, I., and Gottesman, M. M. (1991). Binding of drugs and ATP by P-glycoprotein and transport of drugs by vesicles from human multidrug-resistant cells. *In* "Molecular and Cellular Biology of Multidrug Resistance in Tumor Cells" (I. B. Roninson, ed.), pp. 279–289. Plenum, New York.

Currier, S. J., Ueda, K., Willingham, M. C., Pastan, I., and Gottesman, M. M. (1989). Deletion and insertion mutants of the multidrug transporter. *J. Biol. Chem.* **264**, 14376–14381.

Currier, S. J., Kane, S. E., Willingham, M. C., Cardarelli, C. O., Pastan, I., and Gottesman, M. M. (1992). Identification of residues in the first cytoplasmic loop of P-glycoprotein involved in the function of chimeric human *MDR1-MDR2* transporters. *J. Biol. Chem.* **267**, 25153–25159.

Dalmark, M., and Storm, H. H. (1981). A Fickian diffusion transport process with features of transport catalysis: Doxorubicin transport in human red blood cells. *J. Gen. Physiol.* **78**, 349–364.

Endicott, J. A., and Ling, V. (1989). The biochemistry of P-glycoprotein-mediated multidrug resistance. *Annu. Rev. Biochem.* **58**, 137–171.

Foote, S. J., Kyle, D. E., Martin, R. K., Oduola, A. M. J., Forsyth, K., Kemp, D. J., and Cowman, A. F. (1990). Several alleles of the multidrug-resistance gene are closely linked to chloroquine resistance in *Plasmodium falciparium*. *Nature (London)* **345**, 255–258.

Georges, E., Zhang, J.-T., and Ling, V. (1991). Modulation of ATP and drug binding by monoclonal antibodies against P-glycoprotein. *J. Cell. Physiol.* **148**, 479–484.

Gottesman, M. M., and Pastan, I. (1988). The multidrug-transporter, a double-edged sword. *J. Biol. Chem.* **263**, 12163–12166.

Gottesman, M. M., and Pastan, I. (1993). Biochemistry of multidrug resistance mediated by the multidrug transporter. Annu. Rev. Biochem. **62**, 385–427.

Gottesman, M. M., Schoenlein, P. V., Currier, S. J., Bruggemann, E. P., and Pastan, I. (1991). Biochemical basis for multidrug resistance in cancer. *In* "Biochemical and Molecular Aspects of Selected Cancer" (T. Pretlow and T. Pretlow, eds.), Vol. 1, pp. 339–371. Academic Press, San Diego, California.

Greenberger, L. M., Yang, C. H., Gindin, E., and Horwitz, S. B. (1990). Photoaffinity probes for α-adrenergic receptor and the calcium channel bind to a common domain in P-glycoprotein. *J. Biol. Chem.* **265,** 4394–4401.

Greenberger, L. M., Lisanti, C. J., Silva, J. T., and Horwitz, S. B. (1991). Domain mapping of the photoaffinity drug-binding sites in P-glycoprotein encoded by mouse *mdr*1b. *J. Biol. Chem.* **266,** 20744–20751.

Gros, P., Croop, J., and Housman, D. E. (1986). Mammalian multidrug-resistance gene. Complete cDNA sequence indicates strong homology to bacterial transport proteins. *Cell* **47,** 371–380.

Higgins, C. F., and Gottesman, M. M. (1992). Is the multidrug transporter a flippase? *Trends Biochem. Sci.* **17,** 18–21.

Higgins, C. F., Hiles, I. D., Salmonel, G. P. C., Gill, D. R., Downie, J. A., Evans, I. J., Holland, I. B., Gray, L., Buckel, S. D., Bell, A. W., and Hermodson, M. A. (1986). A family of related ATP-binding subunits coupled to many distinct biological processes in bacteria. *Nature (London)* **323,** 448–450.

Horio, M., Gottesman, M. M., and Pastan, I. (1988). ATP-dependent transport of vinblastine in vesicles human multidrug-resistant cells. *Proc. Natl. Acad. Sci. U.S.A.* **85,** 3580–3584.

Kamijo, K., Taketani, S., Yokota, S., Osumi, T., and Hashimoto, T. (1900). The 70-kDa peroxisomal membrane protein is member of the MDR (P-glycoprotein)-related ATP-binding protein superfamily. *J. Biol. Chem.* **265,** 4534–4540.

Kartner, N., Hanrahan, J., Jensen, T. J., Nalsmith, A. L., Sun, S., Ackerley, C. A., Reyes, E. F., Tsui, L.-C., Rommens, J. M., Benr, C. E., and Riordan, J. R. (1991). Expression of the cystic fibrosis gene in nonepithelial invertebrate cells produces a regulated anion conductase. *Cell* **64,** 681–691.

Keizer, H. G., and Joenje, H. (1989). Increased cytosolic pH in multidrug-resistant human lung tumor cells: Effect of verapamil. *J. Natl. Canc. Inst.* **81,** 706–709.

Kessel, D. (1989). Exploring multidrug resistance using rhodamine 123. *Cancer Commun.* **1,** 145–149.

Kioka, N., Tsubota, J., Kakehi, Y., Komano, T., Gottesman, M. M., Pastan, I., and Ueda, K. (1989). P-glycoprotein gene (*MDR*1) cDNA from human adrenal: normal P-glycoprotein carries Gly[185] with an altered pattern of multidrug resistance. *Biochem. Biophys. Res. Commun.* **162,** 224–231.

Lelong, I. H., Padmanabhan, R., Lovelace, E., Pastan, I., and Gottesman, M. M. (1992). Specificity of the energy requirement of the P-glycoprotein multidrug transporter in plasma membrane vesicles. *FEBS Lett.* **304,** 256–260.

Pastan, I., Gottesman, M. M., Ueda, K., Lovelace, E., Rutherford, A. V., and Willingham, M. C. (1988). A retrovirus carrying an *MDR*1 cDNA confers multidrug resistance and polarized expression of P-glycoprotein in MDCK cells. *Proc. Natl. Acad. Sci. U.S.A.* **85,** 4486–4490.

Raviv, Y., Pollard, H. B., Bruggemann, E. P., Pastan, I., and Gottesman, M. M. (1990). Photosensitized labeling of a functional multidrug transporter in living drug-resistant tumor cells. *J. Biol. Chem.* **265,** 3975–3980.

Richert, N. D., Aldwin, L., Nitecki, D., Gottesman, M. M., and Pastan, I. (1988). Stability and covalent modification of P-glycoprotein in multidrug-resistant KB cells. *Biochemistry* **27,** 7607–7613.

Riordan, J. R., Rommens, J. M., Kerem, B.-S., Alon, N., Rozmahel, R., Grzelczak, Z., Zienlenski, J., Lok, S., Plavsic, N., Chou, J.-L., Drumm, M. L., Iannuzzi, M. C., Collins, F. S., and Tsui, L.-C. (1989). Identification of the cystic fibrosis gene: Cloning and characterization of complementary DNA. *Science* **245**, 1066–1073.

Safa, A. R. (1988). Photoaffinity labeling of the multidrug-resistance-related P-glycoprotein with photoactive analogs of verapamil. *Proc. Natl. Acad Sci. U.S.A.* **85**, 7187–7191.

Safa, A. R., Glover, C. J., Meyers, M. B., Bidler, J. L., and Felsted, R. L. (1986). Vinblastine photoaffinity labeling of a high-molecular-weight surface membrane glycoprotein specific for multidrug-resistant cells. *J. Biol. Chem.* **261**, 6137–6140.

Safa, A. R., Glover, C. J., Sewell, J. L., Meyers, M. B., Biedler, J. L., and Felsted, R. L. (1987). Identification of the multidrug-resistance-related membrane glycoprotein as an acceptor for calcium channel blockers. *J. Biol. Chem.* **262**, 7884–7888.

Safa, A. R., Mehta, N. D., and Agresti, M. (1989). Photoaffinity labeling of P-glycoprotein in multidrug-resistant cells with photoactive analogs of colchicine. *Biochem. Biophys. Res. Commun.* **162**, 1402–1408.

Safa, A. R., Stern, R. K., Choi, K., Agresti, M., Tamai, I., Mehta, N. D., and Roninson, I. B. (1990). Molecular basis of preferential resistance to colchicine in multidrug-resistant human cells conferred by Gly-185 → Val-185 substitution in P-glycoprotein. *Proc. Natl. Acad. Sci. U.S.A.* **87**, 7225–7229.

Schinkel, A. H., and Borst, P. (1991). Multidrug resistance mediated by P-glycoproteins. *Semin. Cancer Biol.* **2**, 213–226.

Stein, W. D., Gottesman, M. M., and Pastan, I. (1993). The multidrug transporter as a protonmotive engine. Submitted for publication.

Tamai, I., and Safa, A. R. (1991). Azidopine noncompetitively inhibits vinblastine and cyclosporin A binding to P-glycoprotein in multidrug resistant cells. *J. Biol. Chem.* **266**, 16796–16800.

Tanaka, S., Currier, S. J., Bruggemann, E. P., Ueda, K., Germann, U. A., Pastan, I., and Gottesman, M. M. (1990). Use of recombinant *p*-glycoprotein fragments to produce antibodies to the multidrug transporter. *Biochem. Biophys. Res. Commun.* **166**, 180–186.

Thiebaut, F., Currier, S. J., Whitaker, J., Haugland, R. P., Gottesman, M. M., Pastan, I., and Willingham, M. C. (1990). Activity of the multidrug transporter results in alkalinization of the cytosol: Measurement of cytosolic pH by microinjection of a pH-sensitive dye. *J. Histochem. Cytochem.* **38**, 685–690.

Valverde, M. A., Diaz, M., Sepulveda, F. V., Gill, D. R., Hyde, S. C., and Higgins, C. F. (1992). Volume-regulated chloride channels associated with the human multidrug resistance P-glycoprotein. *Nature (London)* **355**, 830–833.

Van der Bliek, A. M., Kooiman, P. M., Schneider, C., and Borst, P. (1988). Sequence of *mdr*3 cDNA encoding a human P-glycoprotein. *Gene* **71**, 401–411.

Weaver, J. L., Pine, P. S., Aszalos, A., Schoenlein, P. V., Currier, S. J., Padmanabhan, R., and Gottesman, M. M. (1991). Laser scanning and confocal microscopy of daunorubicin, doxorubicin, and rhodamine 123 in multidrug-resistant cells. *Exp. Cell Res.* **196**, 323–329.

Yoshimura, A., Kuwazuru, Y., Sumizawa, T., Ichikawa, M., Ikeda, S., Ueda, T., and Akiyama, S. (1989). Cytoplasmic orientation and two-domain structure of the multidrug transporter, P-glycoprotein, demonstrated with sequence-specific antibodies. *J. Biol. Chem.* **264**, 16282–16291.

Zamora, J. M., Pearce, H. L., and Beck, W. T. (1988). Physical–chemical properties shared by compounds which moderate multidrug resistance in human leukemia cells. *Mol. Pharmacol.* **33**, 454–462.

CHAPTER 2

A Novel Mechanism for Transmembrane Translocation of Peptides: The *Saccharomyces cerevisiae STE6* Transporter and Export of the Mating Pheromone a-Factor

Karl Kuchler,[1] **Elana E. Swartzman, and Jeremy Thorner**
Department of Molecular and Cell Biology, Division of Biochemistry and Molecular Biology, University of California, Berkeley, California 94720

I. INTRODUCTION

In eukaryotic cells, the vast majority of peptide hormones and other polypeptides destined for secretion are typically synthesized as larger precursors on cytoplasmic polysomes and then targeted to the membrane

---

[1] Current address: Department of Molecular Genetics and the Biocenter, University of Vienna, Dr. Bohrgasse 9/2, A-1030, Vienna, Austria.

of the endoplasmic reticulum (ER) by means of a cleavable hydrophobic segment, called its "signal peptide" (von Heijne, 1985, 1990), which is usually present at (or near) the N-terminus of the polypeptide (Pugsley, 1990). This process requires both a cytosolic factor, signal recognition particle (SRP), as well as receptors on the surface of the ER membrane (Rapoport, 1991). On entry of the nascent polypeptide chain into the lumen of the ER, the signal peptide is usually cleaved (Dev and Ray, 1990). Protein unfolding and refolding, mediated by cytosolic and intraorganellar enzymes of the Hsp70 and Hsp60 class of stress proteins, are involved in maintaining the polypeptides in a "translocation-competent" state during transport and in facilitating the acquisition of their functional state after transport has been achieved (Deshaies *et al.*, 1988; Chirico *et al.*, 1988; Langer and Neupert, 1991). Once properly assembled in the ER lumen, secreted proteins are delivered to their final destination by packaging in transport vesicles. It is the formation ("budding"), movement, and refusion ("targeting") of the vesicles that sorts and conveys secretory proteins from the ER to the Golgi, through the Golgi cisternae, and finally to the plasma membrane (Balch, 1990). Release of proteins from cells can occur continuously ("constitutive secretion"); alternatively, secretory proteins can be stored intracellularly in specialized vesicles, called secretory granules, for subsequent release on an appropriate stimulus ("regulated secretion") (Miller and Moore, 1990).

Although the classical secretory pathway, whose features were summarized above, has traditionally been considered the only route by which peptides and proteins can exit from eukaryotic cells, over the past several years, a number of notable exceptions to this general picture of protein secretion have emerged (Kuchler and Thorner, 1990; Muesch *et al.*, 1990). Eukaryotic cells are able to produce a number of peptides and proteins that lack a typical hydrophobic signal peptide, yet are actively secreted by a mechanism which does not require that the secretory pathway be functional. Our results on the release of the mating pheromone, **a**-factor, by cells of a eukaryotic microorganism, the yeast *Sacharomyces cerevisiae,* and recent findings in other organisms, including humans, suggest that export of at least some proteins is accomplished by the action of dedicated (that is, substrate-specific) translocators. These transporters are integral membrane proteins and derive the energy for driving transmembrane translocation of their substrates from the hydrolysis of ATP. One hallmark common to all of these proteins is that they possess at least one copy of a 150-residue sequence motif (Walker *et al.*, 1992) that folds into a domain that is responsible for the binding and hydrolysis of ATP. This domain has been referred to as the "ATP-binding cassette" or "ABC" domain (Hyde *et al.*, 1990). Related transporters containing the

ABC domain are also found even in bacterial cells (Higgins, 1992). Taken together, these homologous prokaryotic and eukaryotic proteins constitute a superfamily of ubiquitous "traffic ATPases" (Ames *et al.*, 1990) and their structure and domain organization appear to have been highly conserved during evolution.

One representative member of this class of specialized membrane proteins is the *STE6* gene product of the yeast *S. cerevisiae* (Kuchler *et al.*, 1989; McGrath and Varshavsky, 1989). Here we review the evidence that Ste6p is a member of the ABC family of membrane-bound transporters and the findings that demonstrate that Ste6p is responsible for the secretion of the peptide mating pheromone, **a**-factor, from haploid cells of the *MATa* genotype. We also present more recent data on the structure, localization, and function of Ste6p. Using Ste6p as the paradigm, we also relate what we know about the physiological role of this type of transporter to a mammalian homolog of Ste6p, the human *mdr1* gene product or "*P*-glycoprotein," first recognized because of its ability to confer multidrug resistance to animal cells in culture (Endicott and Ling, 1989; Gottesman and Pastan, 1993). Finally, we discuss the clinical relevance of ABC-type transporters in human health and disease.

II. YEAST **a**-FACTOR EXPORT REQUIRES AN ABC-TYPE TRANSPORTER

Although it is a unicellular eukaryotic microbe, *S. cerevisiae* exists in three distinct cell types. There are two haploid cell types, *MATa* and *MATα*, which produce and respond to extracellular peptide hormones. These peptides bind to G-protein-coupled cell surface receptors and initiate a signal transduction pathway that induces processes that permit the haploid cells to conjugate or "mate" to form the third cell type, the *MATa/MATα* diploid cell (for review see Sprague and Thorner, 1992). These peptide hormones are referred to, therefore, as "mating pheromones." *MATα* cells secrete α-factor pheromone, which acts on the *MATa* cells; *MATa* cells produce **a**-factor pheromone, which acts on the *MATα* cells.

Work from this laboratory demonstrated previously that one of the pheromones, α-factor (a tridecapeptide), is generated by *MATα* cells, first, as a 165-residue precursor, which enters the secretory pathway where it undergoes Asn-linked glycosylation and proteolytic processing to produce the mature bioactive peptide prior to its release via secretory vesicles (reviewed in Fuller *et al.*, 1988). Thus, the biosynthesis of α-factor follows the classical secretory pathway. In marked contrast, we discovered more recently that the pheromone produced by *MATa* cells, **a**-factor (a dodeca-peptide), is generated by a completely different route. The **a**-factor is initially synthesized as a 36-residue precursor, but does not enter the secretory pathway and is converted in the cytosol (or at the inner face of the plasma membrane) to its mature form prior to its release (Sterne and Thorner, 1986, 1987; Sterne, 1989). Active **a**-factor carries two post-translational modifications (Anderegg *et al.*, 1988). A farnesyl chain is attached in thioether linkage to the sulfhydryl group of the C-terminal cysteine residue; the genes (*RAM1* and *RAM2*) that encode the subunits of the protein:prenyl transferase responsible for this modification have been identified (Schafer *et al.*, 1989; for additional details see Schafer and Rine, 1992). In addition, the carboxyl group of the C-terminal cysteine is methyl-esterified; the gene (*STE14*) that encodes the S-farnesyl-cysteine carboxyl:S-adenosylmethionine methyltransferase responsible for this modification also has been identified (Sterne-Marr *et al.*, 1990). Thus, mature bioactive **a**-factor is actually a lipopeptide.

Most importantly, we demonstrated (Sterne and Thorner, 1986, 1987; Sterne, 1989; Kuchler *et al.*, 1989) that export of **a**-factor does not require the normal secretory vesicle-based secretion pathway because release of **a**-factor from *MAT***a** cells was not blocked by any of the temperature-sensitive *sec* mutations that define the secretory pathway in *S. cerevisiae* (Schekman, 1985). By examining **a**-factor synthesis in *MAT***a** cells carrying mating-defective or so-called "sterile" (or *ste*) mutations that specifically prevent **a**-factor production by *MAT***a** cells (i.e., which have no effect on α-factor production by *MAT*α cells), we defined gene products that were uniquely required for the generation of bioactive extracellular **a**-factor. In this way, we found that the function of the *STE6* gene product is essential for the release of mature **a**-factor because active **a**-factor accumulates inside *ste6* mutant cells (Kuchler *et al.*, 1989). Determination of the nucleotide sequence of the *STE6* gene showed that it had the capacity to encode a polypeptide that is highly homologous in predicted amino acid sequence and presumptive topology to human Mdr1 and other members of the ABC-transporter superfamily (Kuchler *et al.*, 1989; McGrath and Varshavsky, 1989). Most significantly, we found that overexpression of the *STE6* gene in *MAT***a** cells increased both the rate and extent of **a**-factor secretion dramatically (Kuchler *et al.*, 1989). Thus, the level and activity of Ste6p are rate limiting for the release of **a**-factor by *MAT***a** cells. Given its sequence homology to known transport proteins, like Mdr1 and bacterial permeases, and its requirement in the production of extracellular **a**-factor, we proposed that Ste6p is the membrane protein directly responsible for mediating the transmembrane translocation of the pheromone (Kuchler *et al.*, 1989). We also proposed, by analogy, that the physiological role of Ste6p homologs in other organisms, in particular Mdr1, might be to catalyze the export of peptides and proteins that lack a classical N-terminal hydrophobic leader sequence (Kuchler and Thorner, 1990, 1992a). In fact, to date, Ste6p is the first and only eukaryotic Mdr-like transporter whose authentic and biologically relevant substrate has been unequivocally identified.

III. *STE6* TRANSPORTER IS AN INTEGRAL BUT UNGLYCOSYLATED PLASMA MEMBRANE-ASSOCIATED PROTEIN

The *STE6* gene product belongs, based on sequence homology, to the superfamily of prokaryotic and eukaryotic ATP-dependent transporter proteins. Ste6p and its closest relatives, including human Mdr1 and human CFTR (Riordan *et al.,* 1989), have a distinct domain structure. In the amino-terminal half of these proteins, there are six highly hydrophobic and potentially α-helical segments, each long enough to span a lipid bilayer. These transmembrane segments (TMSs) are followed by the ABC domain that has been shown to contain the ATP-binding motif. The carboxy-terminal half of the molecule contains a nearly identical repeat of the sequences in the amino-terminal portion. Hence, the overall structure of Ste6p, Mdr1, and CFTR could be abbreviated as $[(TMS)_6\text{-}ABC]_2$. A direct role for Ste6p in **a**-factor transport makes several predictions about the topology, membrane insertion, subcellular localization, and regulation of this protein. As one approach for addressing these issues, we generated and employed specific antibodies and other probes to characterize in more detail the structure, membrane association, biochemical properties, and physiological functions of Ste6p.

The *STE6* DNA sequence predicts an open reading frame of 1290 amino acids (calculated molecular weight, 144,744; Kuchler *et al.,* 1989; McGrath and Varshavsky, 1989). To determine if yeast cells actually produce a corresponding polypeptide, and to examine the relationship between the primary translation product and the native functional protein, polyclonal antibodies directed against a TrpE–Ste6 fusion protein were raised in rabbits. When *MAT*a cells were radiolabeled with ^{35}S-Cys and ^{35}S-Met, one of the antisera raised, but not the corresponding preimmune serum, specifically and efficiently immunoprecipitated a protein from total cell extracts solubilized with sodium dodecyl sulfate (SDS) that had an apparent molecular weight of 145,000, in excellent agreement with the size of Ste6p deduced from its gene sequence (Kuchler *et al.,* 1993a). This 145-kDa species was absent in extracts of a *MAT*a *ste6*Δ mutant, and markedly overproduced in *MAT*a cells carrying the *STE6* gene on a multicopy plasmid, confirming that the 145-kDa protein is indeed the product of the *STE6* gene. Because the polyclonal antiserum failed to detect Ste6p on

immunoblots, several epitope-tagged versions of Ste6p were constructed
and employed for detection of Ste6p by immunoblotting. Three epitope-
tagged derivatives of Ste6p were fully functional, as judged by two independent assays for production of extracellular **a**-factor (Kuchler *et al.*,
1993a). One of these derivatives contained a 10-residue antigenic determinant (-EQKLISEEDL-) from the c-Myc oncoprotein (Evan *et al.*, 1985)
inserted in-frame between amino acids 70 and 71 of Ste6p. Another derivative had a longer 16-residue segment (-LEEQKLISEEDLLRKR) containing the same determinant attached in-frame as the C-terminal sequence, immediately following residue 1285 of Ste6p. The third derivative
contained the c-Myc epitopes at both locations. In a similar way, a linker
encoding the so-called ''FLAG'' epitope (-DYKDDDDK-) (Hopp *et al.*,
1988) was inserted in-frame into the *STE6* coding sequence either at a
unique *Eco*RI site (between residues 205 and 206), which lies within the
second predicted internal hydrophilic loop, or at a unique *Bcl*I site (between residues 282 and 283), which lies within the third predicted external
hydrophilic loop. Like all three of the c-Myc-tagged versions, the Ste6p
derivative containing the FLAG epitope in the third external loop was
able to complement the *ste6Δ* mutation for **a**-factor secretion, as judged
by restoration of a halo of growth inhibition in a lawn of a *MATα sst2*
indicator strain, and was capable of promoting mating with an appropriately marked *MATα* partner, a process which also requires production of
extracellular **a**-factor. Thus, in these four derivatives, the presence of the
epitope insertion did not disrupt the proper folding, localization, or function of Ste6p. In contrast, the Ste6p derivative carrying the FLAG insertion within the second internal loop was completely nonfunctional
(Kuchler *et al.*, 1993a).

The presence of even the single 10-residue c-Myc epitope was sufficient
to permit sensitive detection of Ste6p on immunoblots using the anti-
c-Myc monoclonal antibody 9E10 and a chemiluminescence detection
system. For example, the amount of Ste6p expressed from a low-copy-
number (*CEN*) plasmid was readily detectable in 50 μg of protein from a
total cell extract separated by gel electrophoresis (Kuchler *et al.*, 1993a).
For either immunoprecipitation or immunoblotting of Ste6p, it was crucial
that samples solubilized with SDS not be heated above 55°C; when boiled
in the presence of SDS, Ste6p aggregated and did not enter polyacrylamide
gels.

The most homologous counterpart to Ste6p in metazoans, Mdr1, is a
glycoprotein *in situ* (termed ''P-glycoprotein'') that carries a single N-
linked mannose-rich oligosaccharide (Endicott and Ling, 1989). The presumed transmembrane topology of Ste6p closely resembles that of Mdr1
and places 1 of the 14 consensus sites for addition of Asn-linked carbohy-

drate present in Ste6p within the first putative external hydrophilic loop and, hence, presumably exposed to the lumen of the compartments of the secretory pathway. The single Asn-linked chain in Mdr1 is attached to a site in a nearly identical position in this mammalian protein (Endicott and Ling, 1989). However, treatment with tunicamycin, an inhibitor of Asn-linked protein glycosylation, had no detectable effect on the apparent molecular weight of Ste6p, as judged by its mobility on SDS–polyacrylamide gel electrophoresis (PAGE), and solubilized Ste6p failed to bind to concanavalin A–agarose (Kuchler *et al.*, 1993a). Concanavalin A is a lectin that is capable of recognizing both the O- and the N-linked mannose residues that are found attached to authentic yeast glycoproteins (Kukuruzinska *et al.*, 1987). Thus, native Ste6p appears not to carry any oligosaccharide chains and, therefore, glycosylation does not seem to be required for either the membrane delivery or the function of Ste6p (Kuchler *et al.*, 1993a).

Hydropathy analysis of the deduced Ste6 polypeptide predicts 12 highly hydrophobic, potentially α-helical, putatively membrane-spanning segments. To determine whether Ste6p is indeed a membrane-associated protein, subcellular fractionation by differential centrifugation of total cell-free extracts was performed and analyzed by immunoblotting. Ste6p was found exclusively in the membrane fraction; no immunoreactivity was found in the high-speed supernatant fraction nor, as expected, in any subcellular fractions prepared from a *ste6Δ* strain (Kuchler *et al.*, 1993a). To determine the nature of this membrane association, a variety of extraction conditions were employed as a means to solubilize Ste6p from the membrane fraction. Total membranes were suspended in different solutions and resedimented at 200,000*g* in an ultracentrifuge, and equivalent portions of the resulting supernatant solutions were analyzed by immunoblotting. Two treatments that normally release only peripherally bound proteins from membranes (0.1 M Na_2CO_3 at pH 11 and 2 M urea) did not release detectable amounts of Ste6p, as predicted if Ste6p is indeed an intrinsic membrane protein. The nonionic detergent Triton X-100 was nearly as efficient in extracting Ste6p from total membranes as the strong anionic detergent SDS, whereas CHAPSO (a switterionic detergent frequently used for solubilization of intrinsic membrane proteins) was moderately effective in the same circumstances. Thus, as expected for an integral membrane protein, only detergent treatment solubilized Ste6p (Kuchler *et al.*, 1993a).

Although total membranes were isolated in the presence of a serine protease inhibitor and EDTA, these precautions were not sufficient to completely prevent proteolysis of Ste6p during the time required for membrane isolation. When whole cells were suspended in SDS-containing sample buffer prior to lysis with glass beads, proteolytic degradation of Ste6p was virtually eliminated. Much more severe nonspecific breakdown of Ste6p was observed when cells were first converted to spheroplasts in an osmotically stabilizing medium by digestion of the cell wall with Zymolyase at 30°C for 30 min prior to membrane solubilization with SDS. It appears, therefore, that the bulk of the Ste6p in intact spheroplasts was susceptible to the proteases known to contaminate commercial preparations of Zymolyase (Scott and Schekman, 1980), suggesting that Ste6p is located in the plasma membrane and that at least a portion of Ste6p is exposed at the exocellular surface.

Because Ste6p appeared to be an integral membrane protein, and because other mammalian and bacterial ATP-dependent transporters are located in the cytoplasmic membrane, it seems reasonable to presume that Ste6p is localized to the plasma membrane. To test this prediction rigorously, however, the intracellular distribution of Ste6p was analyzed by two independent methods for subcellular fractionation of membranes (Kuchler *et al.*, 1993a). First, the partitioning of Ste6p by differential centrifugation was compared to that of known marker proteins. Just like the plasma membrane-associated H^+-translocating ATPase (*PMA1* gene product), the bulk of the total cellular content of Ste6p sedimented at 12,000g; like Pma1p, the remainder of the total Ste6p sedimented at 100,000g. In marked contrast, none of the phosphoglycerate kinase (*PGK1* gene product) sedimented at 12,000g, whereas the majority of Pgk1p remained in the supernatant fraction even after sedimentation at 100,000g. To demonstrate that the Ste6p found in the 12,000g pellet fraction was membrane-associated rather than simply aggregated or associated with some other particulate material, flotation in a sucrose buoyant density gradient was employed. For this purpose, the material in the 12,000g pellet was suspended in 55% sucrose, overlaid with sucrose solutions of decreasing density, and subjected to prolonged high-speed centrifugation, a method first developed for yeast membrane fractionation by Bowser and Novick (1991). As expected for an authentic membrane-associated protein, Ste6p moved from the dense 55% sucrose cushion to the lighter densities characteristic of membranes. Most significantly, the distribution of Ste6p was identical, fraction-for-fraction, to that observed for the plasma membrane ATPase (Kuchler *et al.*, 1993a).

More than 40 eukaryotic members of the superfamily of Ste6p-like transporters have been described to date. Like the other members of the family that are above 100,000 kDa, Ste6p is predicted from its sequence to have two discrete ABC domains. Hydrolysis of ATP has been shown to energize the transport process mediated by the highly related transporter, Mdr1 (Hamada and Tsuruo, 1988) and by other members of this superfamily (for reviews see Higgins, 1992; Kuchler and Thorner, 1992a). As expected, Ste6p solubilized from membranes with Triton X-100 and partially purified by immunoprecipitation is an ATP-binding protein, as judged by affinity labeling of a 145-kDA species using the photoactivatable ATP analog, 8-azido-ATP (Kuchler *et al.*, 1993a). This cross-linking was specific to Ste6p because no labeled species was detectable in cells not expressing Ste6p, because appearance of the photolabeled 145-kDa product required UV-irradiation, and because labeling was effectively competed by the presence of excess unlabeled nucleotide.

IV. LEVEL AND LOCALIZATION OF *STE6* TRANSPORTER ARE REGULATED

The *STE6* gene is only expressed in *MATa* cells (Wilson and Herskowitz, 1984) and its transcription is significantly induced when *MATa* cells are exposed to α-factor pheromone (K. Kuchler, S. Van Arsdell, R. Freedman, and J. Thorner, unpublished observations). Transcriptional activation by α-factor requires a specific *cis*-acting sequence, the pheromone response element, found upstream of the promotor for *STE6* (Wilson and Herskowitz, 1986) and the promoters for other pheromone-inducible genes (Van Arsdell *et al.*, 1987; Van Arsdell and Thorner, 1987). To determine if mating pheromone induction leads to a concomitant elevation of the level of Ste6p, *MATa* cells expressing epitope-tagged Ste6p from a *CEN* plasmid were treated with α-factor and total extracts were prepared at various times and analyzed by immunoblotting. Treatment of *MATa* cells with α-factor increased the expression of Ste6p by at least an order of magnitude within 100 min after addition of pheromone; in contrast, in the same cells, the level of neither phosphoglycerate kinase nor calmodulin was detectably altered during the same time period (Kuchler *et al.*, 1993a). Increased expression of Ste6p in pheromone-treated cells also resulted in elevated a-factor secretion, consistent with our earlier findings using Ste6p artificially elevated by overproduction from a multicopy vector (Kuchler *et al.*, 1989).

To examine whether Ste6p is a phosphoprotein, and if its phosphorylation is stimulated by α-factor, cells overexpressing Ste6p were radiolabeled with $^{32}PO_4^{2-}$ at steady state in the absence or presence of α-factor, and Ste6p was immunoprecipitated with the rabbit polyclonal anti-Ste6p serum. Even in the absence of pheromone treatment, Ste6p was detectably phosphorylated. In the presence of α-factor, the radioactivity incorporated into Ste6p was significantly increased (Kuchler *et al.*, 1993a); however, the increase in phosphate incorporation mirrored the increase in total

Ste6p observed over the same time period. Therefore, it is likely that increased expression of Ste6p following pheromone induction, rather than increased phosphorylation of preexisting Ste6p molecules, accounts for the observed apparent increase in phosphorylation on α-factor treatment.

As yet another means to determine its intracellular location, the distribution of Ste6p was examined by indirect immunofluorescence. In *MATa* haploid cells expressing epitope-tagged Ste6p from a multicopy plasmid, Ste6p was found throughout the plasma membrane in a patchy, rather nonuniform pattern, as indicated by uneven apparent surface staining (Kuchler *et al.*, 1993a). In addition, fluorescence was consistently observed in vesicle-like bodies that seemed to subtend the inner surface of the plasma membrane. To increase the resolution of this technique, *MATa/MATa* diploid cells expressing the epitope-tagged Ste6p were examined. When viewed by setting the focal plane in the center of these large diploid cells, the patchy rim-like staining and plasma membrane-associated vesicular bodies were more obvious (Kuchler *et al.*, 1993a). Essentially the same pattern of plasma membrane and vesicle-like staining, although less intense, was observed in cells expressing Ste6p from a low-copy *CEN* vector or when optical sections (0.5 μm) of the *MATa/MATa* diploids were observed by confocal microscopy. The staining in vesicle-like bodies could reflect newly synthesized Ste6p within the compartments of the secretory pathway en route to the cell surface. Alternatively, these structures could represent a special pool of secretory vesicles (Holcomb *et al.*, 1988a) containing stored Ste6p, providing a reservoir from which Ste6p could be delivered to the plasma membrane in a rapid (and perhaps regulated) manner.

Given the striking increase in the level of Ste6p that occurred after pheromone stimulation, it was of interest to explore whether localization of Ste6p also was affected by α-factor treatment because cytoskeletal elements and other intracellular structures undergo rather pronounced changes in organization when haploid cells are exposed to pheromone (Babe *et al.*, 1989; Drubin, 1991). As a result of these underlying changes, budding of *MATa* and *MATα* cells is suppressed and the morphology of the cells becomes altered, such that the cell wall and plasma membrane grow asymmetrically to produce a prominent projection. The elongated mating cells fuse at the tips of their respective projections (for review see Sprague and Thorner, 1992). Indeed, a dramatic change in the surface distribution of Ste6p was found on pheromone treatment. After exposure to α-factor, plasma membrane staining of Ste6p either in haploid *MATa* cells or in diploid *MATa/MATa* cells (Kuchler *et al.*, 1993a) was observed primarily along the surface of the projection and most predominantly at its tip. The intracellular vesicle-like structures were present in the pheromone-treated cells, although the number and intensity of staining of these bodies appeared to be reduced, suggesting that the redistribution of Ste6p observed on pheromone treatment may involve, at least in part, mobilization and plasma membrane fusion of the Ste6p-containing vesicles. These vesicles may represent, therefore, an exocytotic compartment previously unidentified in yeast.

V. MAMMALIAN Mdr TRANSPORTERS EXPRESSED IN YEAST CAN MEDIATE a-FACTOR EXPORT

Our finding that Ste6p is the a-factor transporter led us to propose (Kuchler *et al.*, 1989; Kuchler and Thorner, 1990) that other Mdr-like proteins in other systems (reviewed in Kuchler and Thorner, 1992a) might catalyze the transmembrane translocation of peptides or proteins. When a variety of tumor cell types are selected for elevated resistance to any one of a number of lipophilic anticancer drugs, the plasma membrane-associated mammalian Mdr1 transporter (P-glycoprotein) is mutated and amplified (Endicott and Ling, 1989; see also Gottesman, Chapter 1, this volume). Elevated levels of the mutant Mdr1 confer on the cells the capacity to efficiently extrude the anticancer drug as well as a variety of other hydrophobic, but otherwise structurally unrelated, chemotherapeutic agents (including doxorubicin, daunomycin, vinblastine, vincristine, actinomycin D, adriamycin, colchicine, taxol, bleomycin, valinomycin, and gramicidin) (Kane *et al.*, 1990). Hence, such cells display a "multidrug resistance" or Mdr phenotype. Two genes, which are closely linked on the same chromosome in the genomes of mice, hamsters, and humans, encode P-glycoproteins which are ~85% identical at the amino acid sequence level, yet only overexpression of Mdr1 (Chen *et al.*, 1986), and not the other gene (Gros *et al.*, 1988; van der Bliek *et al.*, 1988; Schinkel *et al.*, 1991), confers the multidrug resistance phenotype. Furthermore, the various mammalian P-glycoproteins display differential expression in normal tissues and, in some cases, the expression level is quite high. To date, no function under normal physiological conditions for Mdr1 or the other P-glycoproteins has been established.

It was the observation of striking homology between mammalian Mdr1 and a prokaryotic ABC-type transporter, *Escherichia coli* HlyB, which is responsible for the extrusion of a 110,000 -kDa protein toxin (hemolysin A), that first hinted that ATP-driven transporters might be important in the transmembrane translocation of polypeptides in animal cells (Gerlach *et al.*, 1986). The even closer sequence similarity between Ste6p and mammalian *p*-glycoproteins lent additional strength to the suggestion that the physiological function of at least some ABC-type transporters in animal cells might be to mediate the secretion of peptides and/or proteins by a mechanism distinct from the standard secretory pathway. To test this hypothesis directly, we first used the promoter and translational initiation signals of the *STE6* genes to express in yeast cells a mutant *mdr1* cDNA encoding a multidrug-resistant Mdr1 variant (which carries a G185V mutation) (Kuchler and Thorner, 1992b). We were able to demonstrate that this construct permitted copious expression of mammalian Mdr1 in yeast and that at least a portion of the Mdr1 molecules were inserted into the plasma membrane in a properly folded state, as judged by their ability both to bind ATP and to confer on the yeast cells a moderately elevated level of resistance to the antibiotic valinomycin, an Mdr1-directed drug that is capable of killing yeast (Kuchler *et al.*, 1989). Nonetheless, we could not demonstrate that the drug-resistant version of human Mdr1 was capable of substituting for Ste6p to mediate the export of **a**-factor. Another group, however, subsequently showed independently that the normal form of the murine homolog of human Mdr1, mouse Mdr3, was able to complement a *ste6Δ* mutant, but a multidrug resistance variant of Mdr3 was not (Raymond *et al.*, 1992). Hence, we reexamined whether wild-type human Mdr1 (as opposed to the multidrug-resistant variant we tested initially) is capable of mediating **a**-factor export in place of Ste6p. Indeed, when expressed from a strong constitutive promoter, normal Mdr1, but not the Mdr1(G185V) mutant, was able to mediate a readily detectable level of **a**-factor export, as judged by restoration of mating ability to a *MATa ste6Δ* mutant or by production of a halo of growth inhibition in a *MATα sst2* indicator lawn (Kuchler *et al.*, 1993b). In addition, high-level expression of normal human Mdr1 conferred on the yeast cells nearly complete resistance to the growth-inhibitory effect of even millimolar concentrations of valinomycin. Since the physiologically relevant substrate(s) of Mdr1 are still unknown (West, 1990; Gottesman and Pastan, 1993), our results suggest that the function of P-glycoproteins in mammalian cells could include the transmembrane export of endogenous peptides and/or proteins. Consistent with this view, in a separate and contemporaneous study, it was shown that the Mdr1 homolog in Chinese hamster ovary cells is capable of mediating the efficient extrusion of a growth-inhibitory synthetic peptide, *N*-acetyl-leucyl-leucyl-norleucinal (which blocks the intracellular protease, calpain) (Sharma *et al.*, 1992).

VI. MULTIPLE DEDICATED TRANSPORTERS FOR PEPTIDE EXPORT AND INTERCOMPARTMENTAL PROTEIN TRAFFICKING IN EUKARYOTIC CELLS

The secretion of a growing number of proteins from animal cells does not appear to involve the standard secretory pathway, for example interleukin-1α and interleukin-1β (IL-1β) and both acidic and basic fibroblast growth factors (for review see Kuchler and Thorner, 1992a). Like yeast **a**-factor, these proteins do not have any hydrophobic signal peptide, and they are not glycosylated. Also, like yeast **a**-factor, the mature protein is excised by proteolytic processing from the C-terminal end of a precursor protein and, in some instances, there is evidence that the extracellular form of the polypeptide is posttranslationally (either C-terminally or N-terminally) modified by lipid. Furthermore, extracellular release of these mammalian proteins does not appear to involve the classical secretory pathway. For example, secretion of IL-1β is not inhibited by brefeldin A, a drug which blocks normal secretory transport (Misumi *et al.*, 1986), and the newly made IL-1β is not associated with any membrane-bounded compartment (Singer *et al.*, 1988; Rubartelli *et al.*, 1990). These findings suggest that known Mdr proteins or as yet undiscovered Mdr-like transporters may mediate the release of this novel class of secretory proteins.

In all organisms examined, multiple Mdr-like genes are present (Higgins, 1992; Gottesman and Pastan, 1993). Furthermore, in addition to the class of ABC transporters of the $[(TMS)_6\text{-}ABC]_2$ type represented by Ste6p and Mdr1, there are also members of the superfamily that are essentially half the size, with a domain structure of either $(TMS)_6\text{-}ABC$ or $ABC\text{-}(TMS)_6$. Such "half-sized" ABC transporters in humans include, for example, the *tap1* and *tap2* gene products, which are proteins required for the import of peptide antigens into the endoplasmic reticulum to permit association of such antigens with the Class I major histocampatibility receptors of the human immune system (Monaco, 1992), and Pmp70, a protein found in the membrane of human peroxisomes and which is thought to be necessary for the import of the constituent proteins of this organelle (Gartner *et al.*, 1992). In fact, all of the five known ABC-type transporters in humans are of clinical relevance: (a) CFTR, a $[(TMS)_6\text{-}ABC]_2$ gene product, is defective in cystic fibrosis patients and is thought to be a regulated channel for chloride anion (Riordan *et al.*, 1989); (b) the *tap1* and *tap2* genes, mentioned immediately above, may play a role in autoimmune disease; (c) Pmp70, the peroxisomal membrane protein, when defective, results in a fatal cerebrohepatorenal dysfunction known as Zellweger's syndrome (Gartner *et al.*, 1992); (d) the Mdr1/P-glycoprotein, is responsible, as was discussed in detail above, for a prevalent acquired multidrug resistance that confounds effective chemotherapeutic attacks against various malignancies; and, (e) Mrp1, a newly discovered Mdr-like gene product, appears to be responsible for the acquired drug resistance specifically manifested by lung carcinomas (Cole *et al.*, 1992).

With regard to the diversity of ABC transporters found in a single genome, *S. cerevisiae* is fast outpacing other eukaryotic organisms. The *ADP1* gene was discovered as part of the European effort to sequence chromosome III (Purnelle *et al.*, 1991). The deduced Adp1 polypeptide is most similar in size, sequence, and domain arrangement to the *Drosophila* White (O'Hare *et al.*, 1984) and Brown (Dreesen *et al.*, 1988) proteins, which possess the ABC-(TMS)$_6$ topology and which appear to be involved in the uptake of precursors for the synthesis of the pteridine pigments in the secondary pigment cells of the fly compound eye. Larger Ste6/Mdr1-like transporters have also been identified in *Drosophila* (Wu *et al.*, 1991). Deletions of *ADP1* cause no discernible phenotype and, hence, no function has been ascribed to it. The *S. cerevisiae PDR5* gene was originally identified as a locus on chromosome XV which confers elevated resistance to the protein synthesis inhibitor, cycloheximide, and simultaneous cross-resistance to an unrelated inhibitor, sulfometuron methyl (Leppert *et al.*, 1990). Thus, phenotypically Pdr5p displays one of the hallmarks of an Mdr-like ABC transporter, namely, the ability to confer a multidrug resistance phenotype. Correspondingly, the deduced sequence of Pdr5p indicates that it is a large 1511-residue polypeptide with readily detectable homology to the ABC transporter superfamily (Wang *et al.*, 1992).

As one approach to determine if additional *STE6*-like gene products are encoded in the yeast genome, we utilized polymerase chain reaction (PCR) and primers based on sequences highly conserved in the nucleotide-binding folds of over 30 members of the ABC transporter superfamily, in organisms as evolutionarily distant as yeasts and humans. Among the PCR products we have characterized to date, we have identified at least two novel Ste6p-like genes in the *S. cerevisiae* genome (Göransson and Thorner, 1992; Kuchler *et al.*, 1992). We have designated these genes *SSH1* and *SSH2* (for *S*terile-*S*ix *H*omologs). The sequence of the nucleotide-binding domain of Ssh1p revealed that it is even more homologous to authentic Mdr-like proteins from other organisms than is Ste6p; however, both the length of the open reading frame of the entire gene, which we isolated using the PCR product as a probe, and the size of its transcript, as determined by hybridization to poly(A)$^+$ RNA, indicate that Ssh1p is a half-sized ABC transporter. *SSH1* resides on chromosomes XVI and an *ssh1*Δ mutation has no detectable effect on growth, mating, or sporulation under a wide variety of conditions tested and, hence, its function is unknown.

In contrast, the *SSH2* gene, which also resides on chromosome XVI, appears to be the yeast homolog of the human and rat peroxisomal membrane protein, Pmp70, which has been implicated in the import of peroxisomal proteins (Kamijo *et al.*, 1990; Gartner *et al.*, 1992). Definitive genetic tests are currently underway to determine if *SSH2* function is required for the biogenesis of peroxisomes in *S. cerevisiae*, and reagents are being developed to determine if Ssh2p is a component of the peroxisomal membrane.

A considerable number of additional PCR products were generated with the primers we used (Göransson and Thorner, 1992; Kuchler *et al.*, 1992) and remain to be characterized.

VII. CONCLUSIONS AND PROSPECTUS

Based on the essential role of Ste6p in **a**-factor export (Kuchler *et al.*, 1989) and on its homology to known transport proteins (Kuchler *et al.*, 1989; McGrath and Varshavsky, 1989), we proposed previously that Ste6p is a transporter and that it directly mediates the secretion of **a**-factor by *MAT***a** cells. By analogy to Mdr1 (Endicott and Ling, 1989) and to certain other polytopic membrane proteins (Hartmann *et al.*, 1989), we suggested further that Ste6p is an integral plasma membrane-associated protein containing 12 membrane-spanning segments positioned with both its amino- and its carboxyl-termini on the cytosolic face of the plasma membrane. The effects of insertion of the epitope tags we used for immunological detection provide some indirect support for this topological arrangement. Neither of two different epitopes (c-Myc and FLAG) inserted in regions of the protein predicted to be on the external surface of the cell, nor a c-Myc epitope at the extreme C-terminus, detectably perturbed the function of Ste6p, whereas one of the same epitopes (FLAG) inserted into a region of the protein predicted to face the cytosol was incompatible with the function of Ste6p. That a portion of Ste6p is exposed to the exterior of the cell was further supported by the observation that Ste6p is accessible to attack by the protease(s) present in the enzyme preparation standardly used for the generation of yeast spheroplasts. The application of various labeling and extraction procedures demonstrated that Ste6p is an intrinsic ATP-binding membrane protein with an apparent molecular mass (145 kDa) in excellent agreement with the size of the molecule deduced from its predicted amino acid sequence. Furthermore, both subcellular fractionation and immunofluorescence microscopy demonstrated that Ste6p is located in the plasma membrane.

Unlike both mammalian P-glycoprotein (Endicott and Ling, 1989; Kane *et al.*, 1990) and CFTR (Riordan *et al.*, 1989), yeast Ste6p does not appear to carry any oligosaccharide moieties, even though Ste6p has a consensus site for addition of N-linked carbohydrate at a position precisely analogous to that found in mammalian Mdr1. This observation indicates that glycosylation is required neither for the function of Ste6p function nor for its intracellular transport. In this regard, it is noteworthy that other polytopic plasma membrane proteins of yeast, such as Pma1p, the H^+-translocating ATPase (Serrano *et al.*, 1991), and Fur4p, the uracil permease (Silve *et al.*, 1991), are not N-glycosylated, despite the presence of potential N-glycosylation sites. Likewise, human Mdr1 and mouse Mdr3 are not glycosylated when expressed in yeast in functional form (Kuchler and Thorner, 1992b; Raymond *et al.*, 1992); and, at least for Mdr1, glycosylation is apparently also not essential for its function even in animal cells (Beck and Cirtain, 1982; Ling *et al.*, 1983). In contrast, oligosaccharide addition appears to be important for the proper targeting of CFTR because certain mutations that result in incompletely glycosylated CFTR also lead to degradation of the protein in the ER (Cheng *et al.*, 1990).

Although the majority of Ste6p protein was localized to the cell surface, indirect immunofluorescent staining also detected Ste6p in vesicle-like structures at or near the inner side of the plasma membrane. The spatial distribution of *STE2* gene product, the receptor for α-factor mating pheromone on the surface of *MAT*a cells (Blumer *et al.*, 1988; Reneke *et al.*, 1988), displays almost an identical pattern when examined by indirect immunofluorescence microscopy (Marsh and Herskowitz, 1988; Jackson *et al.*, 1991). The apparent vesicular staining for Ste6p could be an artifact of overexpression; however, this conclusion seems unwarranted because a similar staining pattern, albeit much weaker, was observed when Ste6p was expressed from a *CEN* plasmid instead of from a multicopy vector. The observed vesicle-like bodies could arise from some perturbation of the structure of Ste6p caused by the presence of the c-Myc epitopes, resulting in its mislocalization or aggregation within the secretory compartment. This explanation also seems unlikely because the epitope-tagged Ste6p derivatives examined immunocytologically were fully functional physiologically. Therefore, the vesicular bodies most likely represent newly synthesized Ste6p being delivered to the plasma membrane. In fact, many nascent plasma membrane proteins in yeast can also be found in intracellular vesicles isolated by biochemical fractionation (Holcomb *et al.*, 1988b). On the other hand, it has been reported that a significant fraction of yeast Pma1p exists in an intracellular pool of membrane vesicles in which this H^+-ATPase is fully functional (Serrano *et al.*, 1991). Thus, the vesicles that contain Ste6p may not simply represent way stations in its intracellular transport; rather, the Ste6p-containing vesicles could function as an intracellular reservoir of this transport protein, which in response to an appropriate stimulus (presumably exposure to α-factor) can be mobilized for rapid insertion into the plasma membrane. A precedent for such a regulatory mechanism exists in mammalian adipocytes in the case of insulin-stimulated exocytosis of preexisting vesicles that contain a particular glucose transporter isotype (James *et al.*, 1989). In fact, the regulated exocytotic insertion of plasma membrane transporters and channels appears to be a more general phenomenon (Al-Awqati, 1989).

The mating pheromones (**a**-factor and α-factor) trigger a signal transduction cascade that induces a complex series of biochemical and morphological changes in their corresponding target cells (*MATα* and *MAT***a** haploids, respectively). The responses induced result in a massive rearrangement of the cytoskeleton leading to asymmetric growth and formation of a projection. The tip of this elongated projection is the actual site of conjugation where the two haploid cell types fuse to form a zygote. We found that the level of Ste6p undergoes a dramatic change when yeast cells are exposed to mating pheromone. As we have shown previously, elevated levels of Ste6p result in a dramatic increase in the production of extracellular **a**-factor (Kuchler *et al.*, 1989). Moreover, both of the genes that encode the **a**-factor precursors, *MFa1* and *MAa2*, are also pheromone inducible (Brake *et al.*, 1985; Dolan *et al.*, 1989). Hence, in response to receipt of an α-factor signal, *MAT***a** cells will be provoked into emitting a greatly elevated **a**-factor signal. Furthermore, we also found that the majority of Ste6p in α-factor treated *MAT***a** cells is localized primarily to the tip of the growing projection. Hence, most of the **a**-factor secreted will presumably be released from the tip of this projection, leading to a pronounced anisotropy in the gradient of **a**-factor surrounding the *MAT***a** cell. Our results suggest that, in response to pheromone, a *MAT***a** haploid will release a high level of **a**-factor at a highly localized site on its cell surface, the projection tip. Furthermore, because **a**-factor is a lipopeptide, its hydrophobicity and poor diffusibility will help to maintain its high concentration at the projection tip. Presumably only the *MATα* cell nearest the projection tip of the *MAT***a** cell will be exposed to a concentration of **a**-factor sufficiently potent to elicit a complementary biological response. Thus, Ste6p may be essential for efficient mating by *MAT***a** cells for two distinct reasons: Ste6p is required for **a**-factor secretion per se (Kuchler *et al.*, 1989) and, in addition, Ste6p may be critical for establishment and/ or maintenance of the polarity in **a**-factor secretion that is required for effective courtship and proper partner selection (Kuchler *et al.*, 1993a). This latter role for Ste6p may explain why exogenously added **a**-factor does not rescue efficiently the mating defect of *MAT***a** *Mfa1 mfa2* mutants (Michaelis and Herskowitz, 1988; Marcus *et al.*, 1991). Similarly, during mammalian development, there are often circumstances where one cell induces a differentiative event in a neighboring cell, or a tissue induces developmental changes that are confined to a single layer of overlying epithelial cells. Some of the agents that appear to be responsible for such events, for example IL-1α and IL-1β and acidic and basic fibroblast growth factors, do not appear to be secreted by the classical secretory pathway (Kuchler and Thorner, 1990; Kuchler and Thorner, 1992a), as discussed in detail in a previous section. Given that Ste6p is highly homologous to mammalian Mdr proteins and given that it has been shown that at least one mammalian Mdr protein is capable of transporting heterologous peptides (Raymond *et al.,* 1992; Sharma *et al.,* 1992), the ABC-transporter-mediated release of a differentiative peptide may be a device for the highly localized delivery of a developmental signal that has been evolutionarily conserved from yeast to mammals.

Compelling evidence from a number of divergent systems (reviews in Kuchler and Thorner, 1990, 1992a) indicates that members of the ABC-transporter superfamily are located in a variety of organellar membranes, as well as in the plasma membrane. Indeed, the identification of multiple *STE6*-related genes in *S. cerevisiae,* including the *SSH1* and *SSH2* loci which we have discovered (Göransson and Thorner, 1992; Kuchler *et al.,* 1992), suggests that ABC transporters may be considerably more abundant in eukaryotic cells than first anticipated. The remaining challenge is to devise appropriate methods and approaches to determine the relevant physiological roles of each of these molecules.

Furthermore, for any given ABC-type transporter, a number of questions remain to be resolved with regard to the structure and function of such a molecule. For example, it is clear that Mdr1 resides in the plasma membrane and is able to catalyze the ATP-dependent efflux of at least some of the compounds toward which it confers cellular resistance and, thus, Mdr1 appears to function in certain circumstances like an energy-dependent transporter or pump (Kane *et al.,* 1990). However, as was mentioned above, the true physiologically relevant substrate(s) of Mdr1 are as yet unknown (see West, 1990, for fuller discussion of this issue), and this is the first issue that remains to be resolved. The second outstanding question is what distinguishes the role of Mdr1 from the function(s) of the other members of the Mdr-like protein family. For example, its close homolog, Mdr2, is expressed in a number of different normal human tissue types, yet does not confer multidrug resistance when overexpressed. In contrast, at least one other, less homologous, Mdr-like transporter, Mrp1, does appear to mediate drug resistance (Cole *et al.,* 1992). The third conundrum to be solved is what are the structural determinants of the substrate specificity of Mdr1 and other ABC-type transporters. Since substrate selectivity of Mdr1 clearly can be altered by mutation, there must be some sort of substrate-binding site on the molecule; however, it is difficult to envision a substrate-binding pocket that can accommodate the diverse drugs that Mdr1 pumps, yet is able to discriminate against normal cellular metabolites. A fourth issue that bears on the function of Mdr1 is its native structure. The supramolecular architecture of Mdr1 when it is in its functional state has not been defined to date. It is not clear whether the molecule can function as a monomer in the plasma membrane or whether it must self-associate (or associate with other integral membrane proteins) into higher-order oligomers (dimers, tetramers, etc.). Fifth, it is not known whether efficient assembly or membrane insertion of Mdr1 requires the assistance of cellular factors that facilitate protein folding, such as Hsp70s in either the cytosol or the lumen of the ER. Sixth, it is not known whether any ancillary protein cofactors are required to bind and then deliver to Mdr1 the substrates to be exported. An analogy is the set of low-molecular-weight substrate-binding proteins present in the periplasmic space that hand off substrates to the ABC-type bacterial permeases which import them. Seventh, recent findings suggest that in certain situations, Mdr1 possesses properties and behaviors that are more akin to those displayed by a channel than to those of a pure transporter (Valverde *et al.*, 1992); in this sense, Mdr1 resembles another member of the eukaryotic ABC-protein family, the cystic fibrosis transmembrane conductance regulator or CFTR (Ames and Lecar, 1992).

Given the number of different transport processes operating at any given time in a eukaryotic cell, and given the already widespread occurrence of ABC-type transporters and the wide range of polypeptides that must be translocated from one side of a membrane to another, it should be anticipated that many more members of the superfamily of traffic ATPases and their particular physiological roles remain to be discovered.

Acknowledgments

The work carried out by the authors described in this review was supported by postdoctoral fellowships from the Max Kade Foundation, from the Austrian Chamber of Commerce, and from the Erwin Schrödinger Foundation of the "Fonds zur Förderung der wissenschaftlichen Forschung" of the Austrian government (to K.K.); by NIH Research Grant GM21841; and by funds provided by the W. M. Keck Foundation (to J.T.), and by facilities supplied by the Cancer Research Laboratory of the University of California at Berkeley.

References

Al-Awqati, Q. (1989). Regulation of membrane transport by endocytotic removal and exocytotic insertion of transporters. *In* "Biomembranes, Part S: Transport: Membrane Isolation and Characterization" (S. Fleischer and B. Fleischer, eds.), Methods in Enzymology, Vol. 172, pp. 49–59. Academic Press, San Diego.

Ames, G. F.-L., and Lecar, H. (1992). ATP-dependent bacterial transporters and cystic fibrosis: Analogy between channels and transporters. *FASEB J.* **6**, 2660–2666.

Ames, G. F.-L., Mimura, C., and Shyamala, V. (1990). Bactrial periplasmic permeases belong to a family of transport proteins operating from *Escherichia coli* to human: Traffic ATPases. *FEMS Microbiol. Rev.* **75**, 429–446.

Anderegg, R. J., Betz, R., Carr, S. A., Crabb, J. W., and Duntze, W. (1988). Structure of the *Saccharomyces cerevisiae* mating hormone **a**-factor: Identification of S-farnesyl cysteine as a structural component. *J. Biol. Chem.* **263**, 18236–18240.

Baba, M., Baba, N., Ohsumi, Y., Kanaya, K., and Osumi, M. (1989). Three-dimensional

38   Karl Kuchler *et al.*

analysis of morphogenesis induced by mating pheromone α-factor in *Saccharomyces cerevisiae. J. Cell Sci.* **94**, 207–216.

Balch, W. E. (1990). Molecular dissection of early stages of the eukaryotic secretory pathway. *Curr. Opinion Cell Biol.* **2**, 634–641.

Beck, W. T., and Cirtain, M. C. (1982). Continued expression of Vinca alkaloid resistance by CCRF-CEM cells after treatment with tunicamycin or pronase. *Cancer Res.* **42**, 184–189.

Blumer, K. J., Reneke, J. E., Courchesne, W. E., and Thorner, J. (1988). Functional domains of a peptide hormone receptor: The α-factor receptor (*STE2* gene product) of the yeast *Saccharomyces cerevisiae. Cold Spring Harbor Symp. Quant. Biol.* **53**, 591–603.

Bowser, R., and Novick, P. (1991). Sec15 protein, an essential component of the exocytotic apparatus, is associated with the plasma membrane and with a soluble 19.5 S particle. *J. Cell Biol.* **112**, 1117–1131.

Brake, A. J., Brenner, C., Najarian, R., Laybourn, P., and Merryweather, J. (1985). Structure of genes encoding precursors of the yeast peptide mating pheromone a-factor. *In* "Protein Transport and Secretion" (M.-J. Gething, ed.), pp. 103–108. Cold Spring Harbor Lab. Press, Cold Spring Harbor, New York.

Chen, C., Chin, J. E., Ueda, K., Clark, D. P., Pastan, I., Gottesmann, M. M., and Roninson, I. B. (1986). Internal duplication and homology with bacterial transport proteins in the *mdr1* (P-glycoprotein) gene from multidrug-resistant human cells. *Cell* **47**, 381–389.

Cheng, S. H., Gregory, R. J., Marshall, J., Paul, S., Souza, D. W., White, G. A., O'Riordan, C. R., and Smith, A. E. (1990). Defective intracellular transport and processing of CFTR is the molecular basis of most cystic fibrosis. *Cell* **63**, 827–834.

Chirico, W. J., Waters, G. M., and Blobel, G. (1988). 70K heat shock-related proteins stimulate protein translocation into microsomes. *Nature (London)* **332**, 805–810.

Cole, S. P. C., Bhardwaj, G., Gerlach, J. H., Mackie, J. E., Grant, C. E., Almquist, K. C., Stewart, A. J., Kurz, E. U., Duncan, A. M. V., and Deeley, R. G. (1992). Overexpression of a transporter gene in a multidrug-resistant human lung cancer cell line. *Science* **258**, 1650–1654.

Deshaies, R. J., Koch, B. D., Werner-Washburne, M., Craig, E. A., and Schekman, R. (1988). A subfamily of stress proteins facilitates translocation of secretory and mitochondrial precursor polypeptides. *Nature (London)* **332**, 800–805.

Dev, I. K., and Ray, P. H. (1990). Signal peptidases and signal peptide hydrolases. *J. Bioenerg. Biomembr.* **22**, 271–290.

Dolan, J. W., Kirkman, C., and Fields, S. (1989). The yeast STE12 protein binds to the DNA sequence mediating pheromone induction. *Proc. Natl. Acad. Sci. U.S.A.* **86**, 5703–5707.

Dreesen, T. D., Johnson, D. J., and Henikoff, S. (1988). The brown protein of *Drosophila melanogaster* is similar to the white protein and to components of active transport complexes. *Mol. Cell. Biol.* **8**, 5206–5215.

Drubin, D. (1991). Development of cell polarity in budding yeast. *Cell* **65**, 1093–1096.

Endicott, J. A., and Ling, V. (1989). The biochemistry of P-glycoprotein-mediated multidrug resistance. *Annu. Rev. Biochem.* **58**, 137–171.

Evan, G. I., Lewis, G. K., Ramsay, G., and Bishop, J. M. (1985). Isolation of monoclonal antibodies specific for human *c-myc* proto-oncogene product. *Mol. Cell. Biol.* **5**, 3610–3616.

Fuller, R. S., Sterne, R. E., and Thorner, J. (1988). Enzymes required for yeast prohormone processing. *Annu. Rev. Physiol.* **50**, 345–362.

Gartner, J., Moser, H., and Valle, D. (1992). Mutations in the 70 kDa peroxisomal membrane protein gene in Zellweger syndrome. *Nature Genet.* **1**, 16–23.

Gerlach, J. H., Endicott, J. A., Juranka, P. F., Henderson, G., Sarangi, F., Deuchars, K. L., and Ling, V. (1986). Homology between P-glycoprotein and a bacterial haemolysin transport protein suggests a model for multidrug resistance. *Nature* (*London*) **324**, 485–489.

Göransson, H. M., and Thorner, J. (1992). Eukaryotic Mdr1/P-glycoprotein homologues: Unconventional secretion processes mediated by a growing family of ATP-dependent membrane translocators. *In* "Protein Synthesis and Targeting in Yeast" (M. F. Tuite, J. E. G. McCarthy, F. Sherman, and A. J. P. Brown, eds.), pp. 339–348. Springer-Verlag, Berlin.

Gottesman, M. M., and Pastan, I. (1993). Biochemistry of multidrug resistance mediated by the multidrug transporter. *Annu. Rev. Biochem.* **62**, 385–427.

Gros, P., Raymond, M., Bell, J., and Housman, D. (1988). Cloning and characterization of a second member of the mouse *mdr* gene family. *Mol. Cell. Biol.* **8**, 2770–2778.

Hamada, H., and Tsuruo, T. (1988). Characterization of the ATPase activity of the M_r 170,000 to 180,000 membrane glycoprotein (P-glycoprotein) associated with multidrug resistance in K562/ADM cells. *Cancer Res.* **48**, 4926–4932.

Hartmann, E., Rapoport, T. A., and Lodish, H. (1989). Predicting the orientation of eukaryotic membrane-spanning proteins. *Proc. Natl. Acad. Sci. U.S.A.* **86**, 5786–5790.

Higgins, C. F. (1992). ABC transporters: From microorganisms to man. *Annu. Rev. Cell Biol.* **8**, 67–113.

Holcomb, C. L., Hansen, W. J., Etcheverry, T., and Schekman, R. (1988a). Secretory vesicles externalize the major plasma membrane ATPase in yeast. *J. Cell Biol.* **106**, 641–648.

Holcomb, C. L., Hansen, W., Etcheverry, T., and Schekman, R. (1988b). Plasma membrane protein intermediates are present in the secretory vesicles of yeast. *In* "Molecular Biology of Intracellular Protein Sorting and Organelle Assembly" pp. 153–160. A. R. Liss, New York.

Hopp, T. P., Prickett, K. S., Price, V. L., Libby, R. T., March, C. J., Corretti, D. P., Urdal, D. L., and Conlon, P. J. (1988). A short polypeptide marker sequence useful for recombinant protein identification and purification. *Bio/Technology* **6**, 1204–1210.

Hyde, S. C., Emsley, P., Hartshorn, M. J., Mimmack, M. M., Gileadi, U., Pearce, S. R., Gallagher, M. P., Gill, D. R., Hubbard, R. E., and Higgins, C. F. (1990). Structural model of ATP-binding proteins associated with cystic fibrosis, multidrug resistance and bacterial transport. *Nature* (*London*) **346**, 362–365.

Jackson, C. L., Konopka, J. B., and Hartwell, L. H. (1991). *S. cerevisiae* α pheromone receptors activate a novel signal transduction pathway for mating partner discrimination. *Cell* **67**, 389–402.

James, D. E., Strube, M., and Mueckler, M. (1989). Molecular cloning and characterization of an insulin-regulatable glucose transporter. *Nature* (*London*) **338**, 83–87.

Kamijo, K., Taketani, S., Yokota, S., Osumi, T., and Hashimoto, T. (1990). The 70-kDa peroxisomal membrane protein is a member of the Mdr (P-glycoprotein)-related ATP-binding protein superfamily. *J. Biol. Chem.* **265**, 4534–4540.

Kane, S. E., Pastan, I., and Gottesman, M. M. (1990). Genetic basis of multidrug resistance of tumor cells. *J. Bioenerg. Biomembr.* **22**, 593–618.

Kuchler, K., and Thorner, J. (1990). Membrane translocation of proteins without hydrophobic signal sequences. *Curr. Opinion Cell Biol.* **2**, 617–624.

Kuchler, K., and Thorner, J. (1992a). Secretion of peptides and proteins lacking hydrophobic signal sequences: The role of adenosine triphosphate-driven membrane translocators. *Endocr. Rev.* **13**, 499–514.

Kuchler, K., and Thorner, J. (1992b). Functional expression of human *mdr1* cDNA in *Saccharomyces cerevisiae*. *Proc. Natl. Acad. Sci. U.S.A.* **89**, 2302–2306.

Kuchler, K., Sterne, R. E., and Thorner, J. (1989). *Saccharomyces cerevisae STE6* gene product: a novel pathway for protein export in eukaryotic cells. *EMBO J.* **8**, 3973–3984.

Kuchler, K., Göransson, H. M., Viswanathan, M. N., and Thorner, J. (1992). Dedicated transporters for peptide export and intercompartmental traffic in the yeast *Saccharomyces cerevisiae*. *Cold Spring Harbor Symp. Quant. Biol.* **57**, 579–592.

Kuchler, K., Dohlman, H. G., and Thorner, J. (1993a). The a-factor transporter (*STE6* gene product) and cell polarity in the yeast *Saccharomyces cerevisiae*. *J. Cell Biol.* **120**, 1203–1215.

Kuchler, K., Gottesman, M. M., and Thorner, J. (1993b). In preparation.

Kukuruzinska, M. A., Bergh, M. L. E., and Jackson, B. J. (1987). Protein glycosylation in yeast. *Annu. Rev. Biochem.* **56**, 915–944.

Langer, T., and Neupert, W. (1991). Heat shock proteins Hsp60 and Hsp70: Their roles in folding, assembly and membrane translocation of proteins. *Curr. Top. Microbiol. Immunol.* **167**, 3–30.

Leppert, G., McDevitt, R., Falco, S. C., Van Dyk, T. K., Ficke, M. B., and Golin, J. (1990). Cloning by gene amplification of two loci conferring multiple drug resistance in *Saccharomyces*. *Genetics* **125**, 13–20.

Ling, V., Kartner, N., Sudo, T., Siminovitch, L., and Riordan, J. R. (1983). Multidrug resistance phenotype in Chinese hamster ovary cells. *Cancer Treat. Rep.* **67**, 869–874.

Marcus, S., Caldwell, G. A., Miller, D., Xue, C.-B., Naider, F., and Becker, J. M. (1991). Significance of C-terminal cysteine modifications to the biological activity of the *Saccharomyces cerevisiae* a-factor mating pheromone. *Mol. Cell. Biol.* **11**, 3603–3612.

Marsh, L., and Herskowitz, I. (1988). From membrane to nucleus: The pathway of signal transduction in yeast and its genetic control. *Cold Spring Harbor Symp. Quant. Biol.* **53**, 557–565.

McGrath, J. P., and Varshavsky, A. (1989). The yeast *STE6* gene encodes a homologue of the mammalian multidrug resistance P-glycoprotein. *Nature (London)* **340**, 400–404.

Michaelis, S., and Herskowitz, I. (1988). The a-factor pheromone of *Saccharomyces cerevisiae* is essential for mating. *Mol. Cell. Biol.* **8**, 1309–1318.

Miller, S. G., and Moore, H.-P. (1990). Regulated secretion. *Curr. Opinion Cell Biol.* **2**, 642–647.

Misumi, Y., Misumi, Y., Miki, K., Takatsuki, A., Tamura, G., and Ikehara, Y. (1986). Novel blockade by brefeldin A of intracellular transport of secretory proteins in cultured rat hepatocytes. *J. Biol. Chem.* **261**, 11398–11403.

Monaco, J. J. (1992). A molecular model of MHC class I-restricted antigen processing. *Immunol. Today* **13**, 173–179.

Muesch, A., Hartmann, E., Rohde, K., Rubartelli, A., Sitia, R., and Rapoport, T. A. (1990). A novel pathway for secretory proteins? *Trends Biochem. Sci.* **15**, 86–88.

O'Hare, K., Murphy, C. V., Levis, R., and Rubin, G. M. (1984). DNA sequence of the *white* locus of *Drosophila melanogaster*. *J. Mol. Biol.* **180**, 437–455.

Pugsley, A. P. (1990). Translocation of proteins with signal sequences across membranes. *Curr. Opinion Cell Biol.* **2**, 609–616.

Purnelle, B., Skala, J., and Goffeau, A. (1991). The product of the YCR105 gene located on the chromosome III from *Saccharomyces cerevisiae* presents homologies to ATP-dependent permeases. *Yeast* **7**, 867–872.

Rapoport, T. A. (1991). Protein transport across the endoplasmic reticulum membrane: Facts, models, mysteries. *FASEB J.* **5**, 2792–2798.

Raymond, M., Gros, P., Whiteway, M., and Thomas, D. Y. (1992). Functional complementation of yeast *ste6* by a mammalian multidrug resistance *mdr* gene. *Science* **256**, 232–234.

Reneke, J. E., Blumer, K. J., Courchesne, W. E., and Thorner, J. (1988). The carboxy-terminal segment of the yeast α-factor receptor is a regulatory domain. *Cell* **55**, 221–234.

Riordan, J. R., Rommens, J. M., Kerem, B.-S., Alon, N., Rozmahel, M., Grzelczak, Z., Zielenski, J., Lok, S., Plavsic, N., Chou, J.-L., Drumm, M. L., Ianuzzi, M. C., Collins, F. S., and Tsui, L.-C. (1989). Identification of the cystic fibrosis gene: Cloning and characterization of complementary DNA. *Science* **245**, 1066–1073.

Rubartelli, A., Cozzolino, F., Talio, M., and Sitia, R. (1990). A novel secretory pathway for interleukin-1β, a protein lacking a signal sequence. *EMBO J.* **9**, 1503–1510.

Schafer, W. R., and Rine, J. D. (1992). Protein prenylation: Genes, enzymes, targets, and functions. *Annu. Rev. Genet.* **26**, 209–237.

Schafer, W. R., Kim, R., Sterne, R. E., Thorner, J., Kim, S.-H., and Rine, J. (1989). Genetic and pharmacological suppression of oncogenic mutations in *RAS* genes of yeast and humans. *Science* **245**, 379–385.

Schekman, R. (1985). Protein localization and membrane traffic in yeast. *Annu. Rev. Cell Biol.* **1**, 115–143.

Schinkel, A. H., Roelofs, E. M., and Borst, P. (1991). Characterization of the human Mdr3 p-glycoprotein and its recognition by P-glycoprotein-specific monoclonal antibodies. *Cancer Res.* **51**, 2628–2635.

Scott, J. H., and Schekman, R. (1980). Lyticase: Endoglucanase and protease activities that act together in yeast cell lysis. *J. Bacteriol.* **142**, 414–423.

Serrano, R., Montesinos, C., Roldan, M., Garrido, G., Ferguson, C., Leonard, K., Monk, B. C., Perlin, D. S., and Weiler, E. (1991). Domains of yeast plasma membrane and ATPase-associated protein. *Biochem. Biophys. Acta* **1062**, 157–164.

Sharma, R. C., Inoue, S., Roitelman, J., Schimke, R. T., and Simoni, R. D. (1992). Peptide transport by the multidrug resistance pump. *J. Biol. Chem.* **267**, 5731–5734.

Silve, S., Volland, C., Garnier, C., Jund, R., Chevallier, M. R., and Haguenauer-Tsapis, R. (1991). Membrane insertion of uracil permease, a polytopic yeast plasma membrane protein. *Mol. Cell. Biol.* **11**, 1114–1124.

Singer, I. I., Scott, S., Hall, G. L., Limjuco, G., Chin, J., and Schmidt, J. A. (1988). Interleukin-1β is localized in the cytoplasmic ground substance but is largely absent from the Golgi apparatus and plasma membranes of stimulated human monocytes. *J. Exp. Med.* **167**, 389–407.

Sprague, G. F., Jr., and Thorner, J. (1992). Pheromone response and signal transduction during the mating process of *Saccharomyces cerevisiae. In* ''The Molecular Biology of the *Yeast Saccharomyces*'' (J. R. Broach, J. R. Pringle, and E. W. Jones, 2nd Ed., pp. 657–744. Cold Spring Harbor Lab. Press, Cold Spring Harbor, New York.

Sterne, R. E. (1989). A novel pathway for peptide hormone biogenesis: processing and secretion of the mating pheromone a-factor by *Saccharomyces cerevisiae*. Ph.D. thesis, Univ. of California, Berkeley.

Sterne, R. E., and Thorner, J. (1986). Processing and secretion of a yeast peptide hormone by a novel pathway. *J. Cell Biol.* **103**, 189a.

Sterne, R. E., and Thorner, J. (1987). Peptide hormone export by a novel pathway: Yeast a-factor mating pheromone is a lipopeptide. *J. Cell Biol.* **105**, 80a.

Sterne-Marr, R. E., Blair, L. C., and Thorner, J. (1990). *Saccharomyces cerevisiae STE14* gene is required for COOH-terminal methylation of a-factor mating pheromone. *J. Biol. Chem.* **265**, 20057–20060.

Valverde, M. A., Diaz, M., Sepúlveda, F. V., Gill, D. R., Hyde, S. C., and Higgins, C. F. (1992). Volume-regulated chloride channels associated with the human multi-drug resistance P-glycoprotein. *Nature (London)* **355**, 830–833.

Van Arsdell, S. W., and Thorner, J. (1987). Hormonal regulation of gene expression in yeast. *In* "Transcriptional Control Mechanisms" (D. Granner, M. G. Rosenfeld, and S. Chang, eds.), pp. 325–332. Alan R. Liss, New York.

Van Arsdell, S. W., Stetler, G. L., and Thorner, J. (1987). The yeast repeated element *sigma* contains a hormone-inducible promoter. *Mol. Cell. Biol.* **7,** 749–759.

van der Bliek, A. M., Kooiman, P. M., Schneider, C., and Borst, P. (1988). Sequence of *mdr3* cDNA encoding a human P-glycoprotein. *Gene* **71,** 401–411.

von Heijne, G. (1985). Signal sequences: The limits of variation. *J. Mol. Biol.* **184,** 99–105.

von Heijne, G. (1990). Protein targeting signals. *Curr. Opinion Cell Biol.* **2,** 604–608.

Walker, J. E., Sarate, M., Runswicke, M. J., and Gay, N. J. (1982). Distantly related sequences in the α- and β-subunits of ATP synthase, myosin, kinases, and other ATP requiring enzymes and a common nucleotide binding fold. *EMBO J.* **1,** 945–951.

Wang, M., Balzi, E., Van Dyck, L., Golin, J., and Goffeau, A. (1992). Sequencing of the yeast multidrug resistance *PDR5* gene encoding a putative pump for drug efflux. *Yeast* **8,** S528.

West, I. C. (1990). What determines the substrate specificity of the multi-drug resistance pump? *Trends Biochem. Sci.* **15,** 42–46.

Wilson, K. L., and Herskowitz, I. (1984). Negative regulation of *STE6* gene expression by the α2 product of *Saccharomyces cerevisiae*. *Mol. Cell. Biol.* **4,** 2420–2427.

Wilson, K. L., and Herskowitz, I. (1986). Sequences upstream of the *STE6* gene required for its expression and regulation by the mating type locus in *Saccharomyces cerevisiae*. *Proc. Natl. Acad. Sci. U.S.A.* **83,** 2536–2540.

Wu, C.-T., Budding, M., Griffin, M. S., and Croop, J. M. (1991). Isolation and characterization of *Drosophila* multidrug resistance gene homologs. *Mol. Cell. Biol.* **11,** 3940–3948.

PART II

Structure–Function Relationships in Ion Pumps

CHAPTER 3

Structural Requirements for Subunit Assembly of the Na,K-ATPase

Douglas M. Fambrough, M. Victor Lemas, Kunio Takeyasu,[a] Karen J. Renaud,[b] and Elizabeth M. Inman[c]
Department of Biology, The Johns Hopkins University, Baltimore, Maryland 21218

I. REGULATION OF THE Na,K-ATPase

The Na,K-ATPase plays a central role in physiology (Rossier *et al.*, 1987; Pressley, 1988). In nearly all animal cells it is the transporter principally responsible for the maintenance of high internal potassium ion concentration and relatively low internal sodium concentration. It is also part

[a] Present address: Department of Medical Biochemistry and the Biotechnology Center, The Ohio State University, Columbus, OH 43210.
[b] Present address: Division of Cellular and Molecular Medicine, University of California, San Diego, La Jolla, CA 92093.
[c] Present address: Department of Biological Chemistry, School of Medicine, University of California, Los Angeles, Los Angeles, CA 90024.

Current Topics in Membranes, Volume 41

45

of the set of functions that maintain sodium and potassium ion concentrations of extracellular fluids within narrow ranges. The resulting concentration gradients across the plasma membrane constitute a potential energy source that is tapped for import and export of various other ions and also for import of certain amino acids and sugars. The ion gradients are also essential for "excitable" functions of nerve, muscle, and many other cell types. Levels of expression of the Na,K-ATPase range from a few molecules per square micrometer of cell surface (e.g., in erythrocytes) to a thousand or more molecules per square micrometer in some electrically excitable cells; particularly in polarized epithelial cells, the distribution of Na,K-ATPase molecules in the plasma membrane is also polarized (Morrow *et al.*, 1989; Nelson and Hammerton, 1989). It is not surprising, therefore, that the Na,K-ATPase is regulated in a wide variety of ways in different cell types. This variety is illustrated in Fig. 1, where the various levels of regulation are organized roughly from left to right to correspond to different steps that occur chronologically from transcription of genes encoding the Na,K-ATPase subunits to degradation of Na,K-ATPase molecules in lysosomes: a life history of Na,K-ATPase molecules.

None of the steps in regulation of the Na,K-ATPase has been described in complete molecular detail, but a great deal of evidence has accumulated about some of the underlying mechanisms of regulation. Molecular biological studies have revealed the existence of small sets of genes for α (Shull *et al.*, 1986a; Takeyasu *et al.*, 1990)- and β-subunits of the NaK-ATPase in mammals and birds (Shull *et al.*, 1986b; Takeyasu *et al.*, 1987; Martin-Vasallo *et al.*, 1989; Gloor *et al.*, 1990; Lemas *et al.*, 1991). Each α- and β-isoform is encoded by a separate gene, so the pattern of isoform expression in various cell types is determined largely by gene regulation. Splicing variants have not yet been discovered among mRNAs encoding the subunits, although the related Ca-ATPases have several splice variants. So far there appears to be a one-for-one correspondence between homologous genes in mammals and birds both in the amino acid sequences of the encoded subunits (especially striking for the α-subunits) and in the tissue specificity of expression of each of these genes. Thus, for example, the α1-isoform (93% identical amino acid sequence in birds and mammals) is the predominant α form expressed in kidney and heart, whereas the α2-form is the predominant form in skeletal muscle, and the α3-isoform (95% sequence identity between birds and mammals) occurs predominantly in the brain. Thus, in each cell type a decision must be made as to which isoforms to express.

The conservation of isoform structures in evolution implies that the

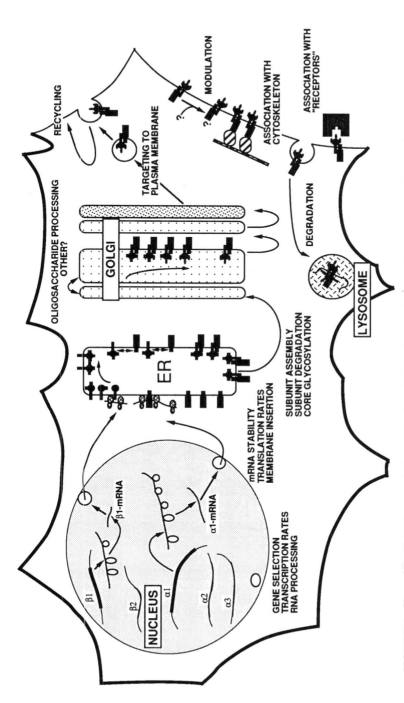

FIGURE 1 Sites of Na,K-ATPase regulation within the cell. The drawing depicts cell compartments directly involved in biosynthesis, function, and turnover of the Na,K-ATPase. Processes known or suspected to be subject to regulation are listed near the relevant compartment.

isoforms serve important, unique purposes in the organism. We have speculated that these differences are particularly in the modes of regulation of function rather than in details of ion transport itself, and this is consistent with recent studies of the functions of individual isoforms (Jewell and Lingrell, 1991; Cone *et al.*, 1991). Although there is a preferential expression of different isoforms in different cell types, there are many examples of cells that express more than one isoform of the α-subunit (Sweander, 1989; Ghosh *et al.*, 1990; McGrail *et al.*, 1991). Such fine analysis of β-subunit expression has been limited up to now by the availability of suitable antibodies for localizing and quantifying β-subunit isoforms in individual cells.

II. SUBUNIT ASSEMBLY AS A CRITICAL STEP FOR REGULATORY CONTROL OF THE Na,K-ATPase

Upregulation of the Na,K-ATPase occurs in a variety of systems in response to perturbations that would be expected to elevate intracellular sodium ion concentration (Lamb and McCall, 1972; Pollack *et al.*, 1981; Pressley, 1988). In studying upregulation of the Na,K-ATPase in tissue-cultured skeletal myotubes, we discovered that subunit assembly plays a critical role in the regulatory process (Wolitzky and Fambrough, 1986; Taormino and Fambrough, 1990). For studies of upregulation we took advantage of skeletal myotubes as a system in which intracellular sodium ion concentration could be increased by pharmacological manipulation of sodium channels in the sarcolemma. Causing these channels to remain opened by including veratridine in the culture medium resulted in a dose-dependent upregulation of the Na,K-ATPase. Closing the channels pharmacologically with tetrodotoxin blocked the upregulatory response and reversed upregulation. The molecular mechanisms underlying upregulation have been examined in studies at the protein and nucleic acid levels. To summarize events in the early phase of upregulation, increased sodium ion entry led to an increase in biosynthesis of sodium pumps sufficient to account quantitatively for the upregulation. The increased biosynthesis was especially marked for the β-subunit and could be accounted for by increased β-gene transcription and selective elevation in the β-mRNA level (Fig. 2). These observations led us to propose that upregulation in this case was due to the production of additional β-subunits, driving the assembly with existing α-subunits to make the extra Na,K-ATPase molecules (see also Lescale-Matys *et al.*, 1990).

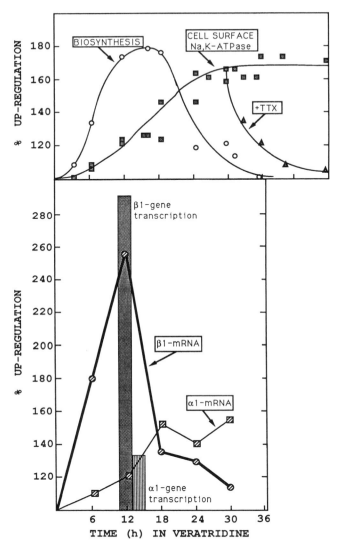

FIGURE 2 Upregulation kinetics for Na,K-ATPase mRNA and protein in primary chick myotubes in response to opening of Na channels with veratridine. Data are compiled from Wolitzky and Fambrough (1986) and Taormino and Fambrough (1990). Peaks of biosynthesis and mRNA expression varied from one experiment to the next. Therefore the data characterize the major events but are not suitable for establishing direct relationships between mRNA levels and biosynthetic rates.

III. ASSEMBLY ASSAY BY COPRECIPITATION WITH SUBUNIT-SPECIFIC MONOCLONAL ANTIBODIES

The minimal functional unit of the Na,K-ATPase appears to be an $\alpha-\beta$ complex (although the form usually occurring in cell membranes may be an $\alpha2\beta2$ complex, and an additional polypeptide called the γ-subunit occurs is some cell types). One of our original interests in studying the Na,K-ATPase was the mechanisms involved in the assembly of subunits. Although the Na,K-ATPase subunits do not interact by covalent bonds, the interaction is strong enough to withstand solubilization in nonionic detergents and also wide variation in the ionic strength of the solvent. Several of our monoclonal antibodies recognize epitopes on the β-subunit that are exposed in the $\alpha-\beta$ complex and thus can be used for immune precipitation of $\alpha-\beta$ complexes. The immune precipitation of β-subunits with α-subunits attached can, therefore, serve as an operational definition of the assembled state (Fig. 3). This assay was used together with pulse-

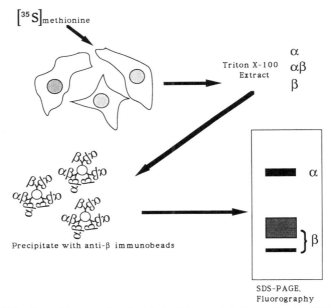

FIGURE 3 Assembly assay for the Na,K-ATPase. Labeling cells with [^{35}S]methionine was used as a procedure to define newly synthesized Na,K-ATPase subunits. Cells were subsequently extracted with detergent solution to solubilize membrane proteins yet preserve existing subunit associations. Precipitation with immunobeads covalently coupled with monoclonal antibody to the avian β-subunit resulted in isolation of free and assembled β-subunits and also the α-subunits present in $\alpha-\beta$ complexes.

chase metabolic labeling to determine that subunit assembly occurs very rapidly during biosynthesis of the Na,K-ATPase and does not depend on β-subunit glycosylation (Fambrough and Bayne, 1983; Tamkun and Fambrough, 1986; see also Takeda *et al.*, 1988). In transfection/expression experiments this assay was used to show that interspecies hybrid α–β complexes assemble in the transfected cells and that certain mutated and chimeric subunits are able to assemble.

IV. KINETICS OF SUBUNIT ASSEMBLY: WHY THEY MAY BE COMPLEX

Under what conditions would selective elevation of β-subunit production lead to assembly of more α–β complexes (Na, K-ATPase molecules)? Clearly, if assembly of subunits were extremely efficient, then the production of Na,K-ATPase molecules would be limited by the supply of the less-abundant subunit. If, on the other hand, assembly of subunits were not efficient, then the production of Na,K-ATPase molecules would be dependent on the supply of each of the two subunits, and elevation of either subunit concentration would lead to increased assembly. In the case of skeletal myotubes in tissue culture in their basal state, then, we would expect to find either (1) a surplus of unassembled α-subunits capable of assembly if β-subunit concentration were elevated or (2) populations of unassembled α- and β-subunits.

A rigorous, quantitative study of the abundance of unassembled α- and β-subunits has never been reported, and this situation is likely due to the enormous technical difficulties that would have to be overcome in order to obtain convincing data. We know that assembly occurs very rapidly during or immediately following biosynthesis of the subunit polypeptide chains (Fig. 4). Pulse–chase metabolic labeling experiments have shown that there is little assembly during the chase period following a short metabolic pulse. Thus, the supply of assembly-competent but unassembled subunits appears to be negligible (Fambrough, 1983). More specifically, (1) the total recoverable ^{35}S-labeled α–β complexes ceased to increase within a few minutes of beginning a metabolic chase following a short pulse label and (2) the ratio of label in the α- and β-subunits remained nearly constant during the labeling period and the chase period. If there had been pools of unassembled α- and β-subunit that could assemble long after their biosynthesis, the incorporation of radioactivity into α–β complexes would have continued to increase during the chase period, which did not happen. If there had been a pool of one subunit but not the other capable of assembly long after biosynthesis, then the ratio of label in the two subunits of immune precipitated α–β complexes would have

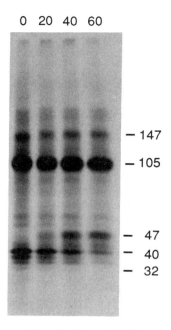

FIGURE 4   Kinetics of assembly and oligosaccharide processing of the Na,K-ATPase. After a 10-min labeling period with [^{35}S]methionine, primary chick myotubes were cultured in medium lacking [^{35}S]methionine for 0, 20, 40, or 60 min. Cells were then extracted with detergent solution, and the β-subunits were isolated with immunobeads. Immunoprecipitates were analyzed by sodium dodecyl sulfate–polyacrylamide gel electrophoresis (SDS-PAGE) and fluorography. α-Subunit migrated with M_r, 105 kDa; mature β-subunit, 47 kDa; and high-mannose intermediate β-subunit forms between 36 and 40 kDa. Note lack of accumulation of more radioactivity in the subunits during the chase period and the occurrence of a set of high-mannose glycosylation intermediates of the β-subunit at the start of the chase period.

drifted with time as more and more of the subunits from the pool were incorporated into α–β complexes. From this study it was argued that not only is subunit assembly a rapid process that occurs during or immediately following subunit biosynthesis, but any residual unassembled subunits must lose their competence for assembly and/or be degraded very rapidly. Measurements of what was interpreted to be a pool of unassembled β-subunits revealed rather rapid degradation (over a time course of about 1 hr).

There are two questions here: (1) how large are the populations of unassembled subunits? and (2) what fractions of these unassembled subunit populations are competent for assembly? As more and more is learned about membrane protein biosynthesis, a general picture has emerged.

There is substantial degradation of newly synthesized molecules, amounting in the case of acetylcholinesterase to about 90% of the molecules synthesized (Rotundo, 1988). Unassembled subunits are generally rapidly degraded in the endoplasmic reticulum (ER) (most data coming from experiments in which individual subunits were expressed separately in host cells that did not express the whole functional multisubunit complexes) (Bonifacino *et al.*, 1990a). Assembled subunit complexes are transported out of the ER to the Golgi apparatus fairly rapidly (judged in most cases by the rate of formation of complex oligosaccharides on glycoprotein subunits). Given this picture, we expect the populations of unassembled subunits to be very modest (not amounting to more than 30–60 min worth of new synthesis), to be located in the ER, and to be degraded by a fast but not necessarily kinetically simple mechanism. Even if we could measure the sizes of the unassembled populations of subunits in the ER, there remains the question of whether these are still assembly-competent subunits. The unassembled molecules might be folded incorrectly and so not recognize one another. There is some evidence that unassembled subunits interact with heat-shock-related binding protein (BIP) (Bole *et al.*, 1986) in the ER, which might prevent assembly (see below). Other mechanisms might sequester unassembled subunits into regions of the ER

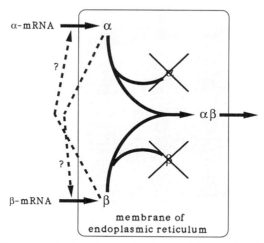

FIGURE 5 Kinetic complexity of α–β subunit assembly. The drawing depicts some of the events in the endoplasmic reticulum that relate to Na,K-ATPase biosynthesis. Subunits are synthesized from separate mRNAs but there may be influences of one subunit on the biosynthesis of the other. Assembly and subunit degradation are shown as alternate pathways, although there may also be degradation of assembled subunits so that only a subset of initial α–β complexes exit the ER en route to the plasma membrane.

in which assembly is disfavored or does not occur at all. Given all these possibilities, measuring the assembly competence of unassembled subunits might be beyond the capability of current technology.

In addition to all the possibilities mentioned above, it is possible that elevated expression of the β-subunit in skeletal myotubes might somehow stimulate increased translation of α-subunit mRNA. In this case, there might never be a pool of unassembled α-subunits but rather a production rate of α-subunits proportional to the number of β-subunits synthesized. One can imagine a variety of mechanisms to achieve this, including direct interactions of nascent β-subunits or newly synthesized β-subunits with α-subunit mRNAs or nascent α-subunits to stimulate α-subunit mRNA translation. Finally, it remains possible that in the upregulation of Na,K-ATPase in skeletal myotubes there is independent stimulation of β-subunit gene expression and of α-subunit mRNA translation to yield elevated supplies of both subunits and thereby drive the formation of extra α–β complexes. Many of the ideas in this section are summarized in drawing form in Fig. 5. Whatever the final explanation of the molecular basis of upregulation in skeletal myotubes, the assembly of subunits to form new Na,K-ATPase molecules is an essential aspect of the process.

V. TRANSFECTION EXPERIMENTS FOR STUDIES OF SUBUNIT ASSEMBLY

As part of our characterization of cDNAs encoding the avian α- and β-subunits of the Na,K-ATPase mammalian cells were transfected with avian cDNAs, singly or in combinations, to study the assembly process (Takeyasu *et al.*, 1987, 1988, 1989; Renaud *et al.*, 1991). When only one avian subunit was expressed in mammalian cells, α–β complexes were recovered that were interspecies hybrids between the avian and mammalian subunits. When both avian subunits were expressed, avian Na,K-ATPase molecules as well as interspecies α–β complexes could be recovered from the mammalian cells. This made it possible to explore the aspects of subunit structure necessary and sufficient for subunit assembly, and some of our findings are summarized below.

In our studies of avian Na,K-ATPase subunits expressed in mammalian cells we also looked at the distribution of the molecules in the cells. The results suggested a role for the β-subunit in Na,K-ATPase production. When avian α-subunit alone was expressed in mouse cells, some of this avian α-subunit arrived at the mouse cell surface and contributed a ouabain-sensitive potassium uptake mechanism. This was an anticipated property of Na,K-ATPase molecules that were hybrids between chicken

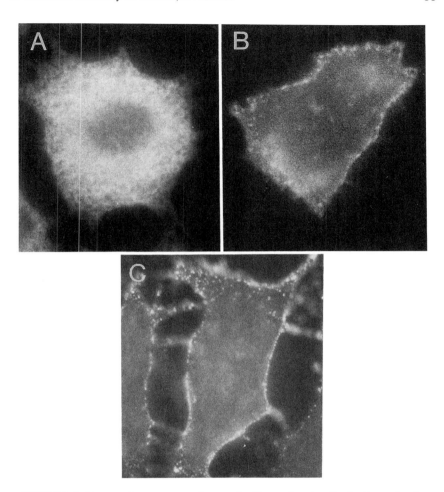

FIGURE 6 Immunofluorescence micrographs illustrating the distributions of avian α (A)- and β (B)-subunit in mouse cells expressing single avian Na,K-ATPase subunits and distribution of avian α-subunits (C) in mouse cell expressing both avian α- and β-subunits. The fluorescence pattern in A is indicative of predominantly endoplasmic reticulum localization, whereas the patterns shown in B and C are indicative of predominantly plasma membrane localization. In all cases cells were permeabilized prior to immune labeling so that the entire populations of subunits were labeled by the fluorescent antibodies.

α- and mouse β-subunit, since the α-subunit was expected to carry the ouabain-binding site. This has now been studied in considerable detail by Price and Lingrel (1988). An observation that was not anticipated was the finding of most of the avian α-subunits in the endoplasmic reticulum of the mouse cells. We interpreted this as indicating that the α-subunits were being produced in excess of what could assemble with mouse β-subunits and be transported to the plasma membrane (i.e., assembly is a prerequisite for plasma membrane expression) (see also Noguchi *et al.*, 1987; McDonough *et al.*, 1990). Hence the β-subunit plays a key role in Na,K-ATPase expression by being necessary for the exit of α–β complexes from the endoplasmic reticulum. In support of this, when avian β-subunit was expressed in mammalian cell lines already expressing avian α, the coexpressing cells generally had high levels of expression of the avian sodium pumps on the plasma membrane and relatively little retention of the α-subunit in the ER (Takeyasu *et al.*, 1989) (Fig. 6). As a few other multisubunit plasma membrane proteins have been studied, it seems to be generally true that the assembly of subunits is a prerequisite for expression at the cell surface. Studies on the acetylcholine receptor and on T-cell receptors are examples.

VI. TOPOLOGY MODELS AND INSIGHTS INTO ASSEMBLY REQUIREMENTS

The number of membrane-spanning regions in the α-subunit of the Na,K-ATPase has been difficult to deduce. Hydropathy profiles suggested anywhere from 6 to 10, and the situation remains unresolved. Currently we favor a 10-span model (Fig. 7), principally based on the following line of reasoning. The Na,K-ATPase and the Ca-ATPase are members of the gene family of E1E2 ATPases, sharing significant sequence homology and a common overall mechanism of cation transport. The hydropathy plots of Na,K-ATPase α-subunits and the sarcoplasmic reticulum Ca-ATPases are quite similar. The proposed membrane-spanning regions of the Ca-ATPase include some negatively charged residues shown by site-directed mutagenesis to be essential for cation recognition (Clarke *et al.*, 1989); these sites occur also in hydrophobic regions of other E1E2 ATPases, including the Na,K-ATPase. Both amino- and carboxyl-termini of the Ca-ATPases (SERCA1 and SERCA2a) are known to be on the cytosolic side of the SR membrane (Matthews *et al.*, 1990; Campbell *et al.*, 1992). An epitope in a hydrophilic region of the Ca-ATPase SERCA1 has been located on the lumenal side of the SR membrane, imposing significant restrictions on models for the transmembrane topology of the Ca-ATPase (Matthews *et al.*, 1990; Clarke *et al.*, 1990), rendering a 10-membrane span model most likely. Given the constraints on possible topological models of the Ca-ATPase and the degree of similarity in sequence, hydropathic character, and mechanisms of function (Jorgensen and Ander-

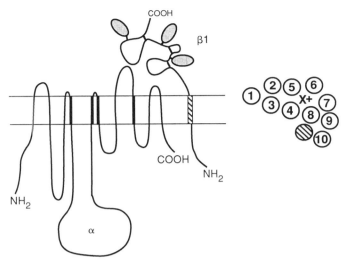

FIGURE 7 Membrane topologies of the Na,K-ATPase subunits. The drawing at left illustrates the predicted course of the α- and β-subunit polypeptide chains through the plasma membrane lipid bilayer. The β1-subunit is illustrated with its three disulfide bonds and three N-linked oligosaccharide groups. At right is a drawing of a possible relationship among the transmembrane segments of the α- and β-subunits. A cation is symbolized in the region surrounded by α-subunit transmembrane segments 4,5,6, and 8, in accordance with the findings of Clarke *et al.* (1989).

sen, 1988), we expect the Na,K-ATPase α-subunit and the Ca-ATPase to have the same topology.

There is no longer significant controversy about the topology of the β-subunit (Fig. 7). A single hydrophobic region has been shown to be necessary and sufficient for membrane insertion of the β-subunit, and the pattern of disulfide bonds and oligosaccharide additions in the extracellular domain leave no possibilities for additional membrane-spanning regions (Kirley, 1989).

The reason for discussing topologies is that these models of structure, crude though they be, constitute a large part of the basis for making decisions about what to do when taking a molecular biological approach to the problem of subunit assembly. This approach begins with a consideration of which regions of each subunit might interact with the other subunit in the α–β complex; subunit topologies set major constraints on which regions can be touching in the complex. Figure 7 illustrates the subunits in the membrane. From such a drawing one can immediately see that any intersubunit interaction that occurs on the cytosolic side of the plasma membrane must involve the N-terminal region of the β-subunit; any interaction within the membrane must involve the single-membrane span of the β-subunit, and interactions on the external side of the plasma membrane must involve the single ectodomain of the β-subunit and one or more of the protruding loops of the α-subunit.

VII. β-SUBUNIT FEATURES REQUIRED FOR ASSEMBLY WITH α-SUBUNITS

To explore the regions of the β-subunit involved in subunit interactions (Renaud *et al.*, 1991), we expressed DNA encoding the avian β-subunit in mouse cells, pulse-labeled the cells with [^{35}S]methionine, solubilized membrane proteins in a buffer containing Triton X-100, and isolated avian β-subunits by immune precipitation with an avian-specific monoclonal antibody. The ^{35}S-labeled, affinity-purified material was analyzed by sodium dodecyl sulfate–polyacrylamide gel electrophoresis (SDS–PAGE) and autoradiography. If subunit assembly had occurred, mouse α-subunit would be seen in the autoradiographs as a band of 100 kDa. In such experiments the formation of α–β complexes was so efficient that approximately stoichiometric amounts of α- and β-subunits were isolated (Fig. 8). Since the methionine content of the α-subunit is about seven times that of the β-subunit, the mouse α-subunit is far more prominent than the α subunit in the autoradiograph.

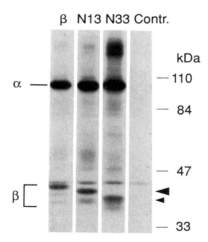

FIGURE 8 Assembly of N-terminal-truncated avian β-subunits with mouse α-subunits: SDS–PAGE fluorographic analysis of immune precipitates prepared as illustrated in Fig. 3 from mouse L-cells transfected with the avian β-subunit cDNAs. (After Renaud *et al.*, 1991.)

We fully expected that subunit association would involve the cyto-plasmic domains of the two subunits. The cytoplasmic domain of the β-subunit is very well conserved in evolution, as expected for a region that must interact with another polypeptide that is, itself, highly conserved in evolution. Therefore we constructed DNAs encoding N-terminally trun-cated forms of the β-subunit and expressed these in mouse cells. By our assembly assay these were shown to assembly with the mouse α-subunit with avidity similar to that of the full-length avian β-subunit (Fig. 8), even thought the more extreme truncation removed the entire cytoplasmic domain of the β-subunit. Immunofluorescence microscopy revealed that these truncated β-subunits were targeted to the plasma membrane.

The single transmembrane domain of the β-subunit is also highly con-served in evolution. If the membrane span were deleted from the β-subunit, it would not integrate into a membrane and so could not assemble with α-subunits. Therefore, to test the role of the membrane-spanning region in assembly, we used a series of partial deletions of this region that, in sum, deleted all but 6 of the 28 amino acids in the hydrophobic region. These deletions were first made in the human β-subunit DNA by Kawakami, who used synthetic mRNAs made from these DNAs to explore the role of the membrane span in integration of the β-subunit into micro-somal membranes during *in vitro* translation (Kawakami and Nagano, 1988). To use these deletion mutants in our assembly assay we needed to preserve the avian β-subunit epitope recognized by the monoclonal antibody. Thus we constructed chimeric β-subunits that included approxi-mately the first 300 bases of the human β-subunit cDNA spliced in-frame with the avian DNA encoding most of the ectodomain of the β-subunit. The chimera with a nonmutated human component assembled well with mouse α-subunits when its DNA was transfected into mouse cells. In addition, all of the chimeric deletion forms that retained the ability to integrate into membranes also retained the ability to assemble with α-subunits, although the efficiency of assembly appeared to be reduced greatly for the more extreme deletions. Deletions of more than 3 amino acids of the β-subunit membrane-spanning domain resulted in retention of the α–β complexes in the endoplasmic reticulum. These were the only cases we have found in which assembly and exit from the endoplasmic reticulum were not associated characteristics. The behavior of the set of membrane span deletions, taken together, suggested that no region of the membrane-spanning domain was absolutely necessary for subunit as-sembly.

The results with the N-terminal truncations and the membrane span deletions suggested that the important sites of α–β interaction might lie outside the plasma membrane. Therefore, a series of C-terminal truncations of the β-subunit were engineered. Subunit assembly was greatly inhibited or prevented by removal of only a few amino acids from the C-terminus of the β-subunit (Fig. 9). The smallest truncation that grossly affected assembly was only 4 amino acyl residues, and truncations of 11 and 19 residues similarly failed to assembly efficiently. In immune precipitates of these expressed β-subunits, only traces of mouse α-subunit were seen. Instead, significant amounts of another polypeptide coprecipitated with the truncated β-subunits. This polypeptide comigrated with the heat-shock cognate protein BIP in SDS–PAGE, and immune precipitates of BIP with a monoclonal antibody also contained truncated avian β-subunit. Much of the truncated β-subunit could not be precipitated with monoclonal antibodies that recognized conformations of native β-subunit but was precipitated with a monoclonal antibody that recognized denatured (but not native) β-subunits. These results suggested that the C-terminus might be very important in α–β assembly, but the experiments

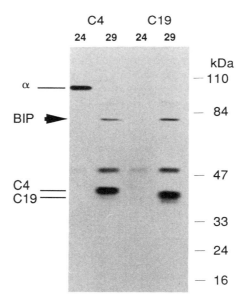

FIGURE 9   Assembly of C-terminal-truncated avian β-subunits with mouse α-subunits: SDS–PAGE fluorographic analysis of immune precipitates prepared as illustrated in Fig. 3 from mouse L-cells transfected with the avian β-subunit cDNAs. Immunobeads used for the immune precipitations were coupled with a monoclonal antibody recognizing native β-subunits (mAb 24) or misfolded β-subunits (mAb 29). C4 and C19 refer to the C-terminal deletions or 4 and 19 amino acids, respectively. The arrowhead marks the position of BIP in the SDS–PAGE fluorographs. Note that C4 assembles with the mouse α-subunit better than does C19, but the majority of β-subunits are unassembled in both cases.

can be interpreted in another way. It may be that C-terminal truncation interferes with proper folding of the β-subunit, resulting in fairly stable association with BIP. This association or the failure to fold correctly, rather than lack of the assembly domain, might explain the failure of $\alpha-\beta$ assembly. In addition, there are normal mouse β-subunits assembling with mouse α-subunits in the transfected cells. If the truncated avian β-subunits are compromised in their ability to assemble, then they may simply be out-competed by mouse β-subunits. Indeed, the fact that traces of α-subunit were found in immune precipitates of C-terminally truncated β-subunits argues for the occurrence of the necessary sequences for assembly in the truncated β-subunits.

We have not completely resolved the issue of assembly of C-terminally truncated β-subunits with α-subunits. Our current approach is to express the truncated β-subunit forms in mouse cells overexpressing chicken α-subunit. Under this condition, the truncated β-subunits should not have to compete with endogenous mouse β-subunits for a limited supply of α-subunit. And if there is any species selectivity in the formation of $\alpha-\beta$ complexes, this situation should also favor formation of avian $\alpha-\beta$ complexes in the transfected cells. An additional approach is to set up conditions disfavoring the interactions of BIP with nascent polypeptide chains. Recently it has been shown that BIP interactions with misfolded proteins depend on the high calcium ion concentration in the ER and can be disturbed by inhibition of the ER Ca-ATPase with thapsigargin (Sagara and Inesi, 1991; Bonifacino *et al.,* 1990b). Therefore, we are examining the possibility that in thapsigargin-treated cells truncated β-subunit interactions with BIP will be destabilized and interactions with the α-subunit therefore allowed.

Taken together, (Fig. 10) the assembly experiments with β-subunit mutants implicate only the external regions of α- and β-subunits in assembly interactions. If this were so, then one might predict that a mutant form of β-subunit that lacked both the N-terminus and the membrane span would, nevertheless, assemble with α-subunits, if it had access to the lumen of the ER. We are testing this prediction with a "secretory" β-subunit expressed from a cDNA constructed by ligating the signal-peptide-encoding region of LEP100 cDNA (Fambrough *et al.,* 1988) in-frame to the encoding DNA for the ectodomain of the β-subunit.

VIII. CHIMERAS FOR STUDIES OF ASSEMBLY REQUIREMENTS OF THE α-SUBUNIT

The studies on the structural requirements in the β-subunit for subunit assembly illustrate the fact that we gain the most information from subunit forms that do assemble. In the cases where assembly is prevented, we remain uncertain whether to ascribe the deficit to loss of a critical region for assembly or to secondary phenomena such as improper folding of the subunit, interaction with BIP, or metabolic destabilization. One approach that attempts to avoid these secondary problems is construction of chimeric proteins for the catalytic subunit, using parts from catalytic subunits of different E1E2 ATPases. This approach is exemplified in the volume in the chapter by Caplan and co-workers, who are making chimeras between the H,K-ATPase and Na,K-ATPase α-subunits and studying their behavior in polarized epithelia. We have also used the chimera approach to examine subunit assembly, but our chimeras involve the Na,K-ATPase α-subunit and the SR/ER-type Ca-ATPases. The Ca-ATPases lack a β-type subunit. They are expressed in cells that also express the Na,K-ATPase, and no aberrant association of the Na,K-ATPase β-subunit with the Ca-ATPase has ever been seen. Between these there is about 25% amino acid identity, largely clustered in certain regions of the molecules. In constructing the chimeras, we have chosen, where possible, fusion points that lie in regions of high sequence homology. Therefore, at the regions of junction between Ca and Na,K-ATPase sequence, the higher-order structure might be preserved and be common to the two-parent forms.

Our first chimeras (Luckie *et al.*, 1991, 1992) resulted from splicing together the encoding DNAs at a common *Eco*N1 site that lies at the position encoding amino acid 715 in the Na,K-ATPase. The resulting chimera that contains the N-terminal two-thirds of the avian SERCA1 Ca-ATPase (Karin *et al.*, 1989) we refer to as CCN, and the complementary chimera that contains the N-terminal two-thirds of the avian Na,K-ATPase α1-subunit (Takeyasu *et al.*, 1987) we call NNC. DNAs encoding these and other chimeras were transfected into mouse cells expressing the avian β-subunit. Then the assembly assay basically as described above was used to isolate the avian β-subunit and any associated catalytic subunits. The only additional step was to distinguish the mouse α-subunit from the chimeric catalytic subunit in the immune precipitates of assembled avian β-subunit. For chimeras CCN and NNC the electrophoretic mobilities differed slightly from that of mouse α-subunit, simplifying analysis.

Of the original two chimeras, CCN assembled with avian β-subunits and appeared to be targeted to the cell surface (Fig. 11). Since the Ca-ATPase itself lacks any association with the Na,K-ATPase β-subunit, we conclude that the region of α-subunit in the chimera should include the β-subunit interaction region. This result with the CCN chimera greatly reduces the regions of the α-subunit to examine for β-interaction. Given that the interaction site is likely to be external to the lipid bilayer of the

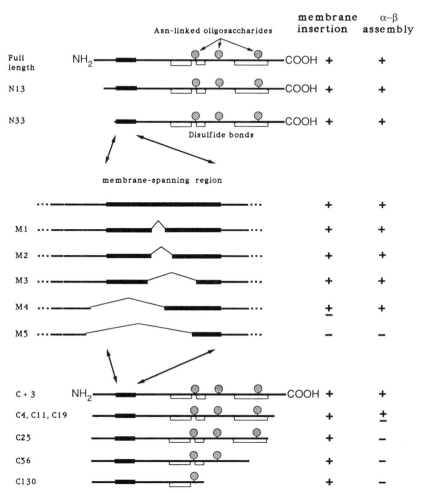

FIGURE 10   Summary diagram of β-subunit mutations, their integration into membrane of the endoplasmic reticulum, and their assembly with mouse α-subunits.

plasma membrane (discussed above), we now begin to focus on the three external loops of the α-subunit, H5H6, H7H8, and H9H10, as candidate regions for β-subunit interaction. Here we come face to face with the topology problem again. We do not precisely know where these supposed loops occur in the primary sequence or even if all three of these actually exist. Amino acid sequence homologies between Ca- and Na,K-ATPases are extremely poor in the C-terminal third of the catalytic subunit, so our

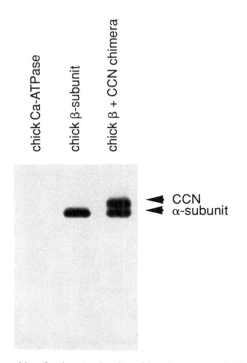

FIGURE 11   Assembly of avian β-subunits with avian α-subunit/Ca-ATPase chimeras in mouse L-cells. SDS–PAGE fluorographic analysis of immune precipitates prepared as illustrated in Fig. 3 from mouse L-cells transfected with the avian cDNAs. Left lane, cells transfected with avian Ca-ATPase cDNA but not β-subunit (one of several controls in these experiments). Center lane, mouse α-subunit precipitated with anti-avian β-subunit immunobeads from cell expressing only avian β-subunit, showing interspecies assembly. Right lane, mouse α-subunit and avian CCn chimera precipitatd with avian β-subunit immunobeads from mouse cells expressing both avian β-subunit and chimera CNN.

rule for making the chimeric junctions within regions of sequence identity cannot always be followed. A set of chimeras with smaller and smaller regions from the Na,K-ATPase has been constructed, and these may be useful in defining the region of α–β interaction further.

In defining the α–β interaction region of the α-subunit, we are, in effect, obtaining a partial solution to the topology problem, since the interactive site(s) should be on the external face of the plasma membrane. The region of the Ca-ATPase defined by polyclonal and monoclonal antibodies as being on the lumenal side of the SR seems to have a counterpart in the Na,K-ATPase that would fit the criterion of being topologically outside the cell. This is an interesting region of the α-subunit, because it is one of the few regions within the C-terminal third of the α-subunit that contains a number of α-isoform-specific residues (Takeyasu *et al.*, 1990). If this region participates in interactions with the β-subunit, then the α–β interactions might be expected to show isoform specificity.

IX. THE ISOFORMS ISSUE IN ASSEMBLY

Can each α-subunit isoform interact with each β-subunit isoform? This question has been addressed indirectly by coexpressing various α-subunit isoforms and β-subunit isoforms in *Xenopus* oocytes and scoring Na,K-ATPase activity. Such studies suggest that various combinations of isoforms can form functional Na,K-ATPase molecules. The subunit assembly of the three avian α-subunit isoforms with the β1-isoform has been demonstrated directly. In these experiments we transfected the β1-subunit DNA into mouse cell line expressing each of the avian α-subunits. α–β complexes were immune precipitated with an anti-avian β-subunit mAb. The coprecipitated α-subunits were dissociated from the immune complexes and reprecipitated with a mAb specific for the avian α-subunits in this system. Each of the three avian α-subunits was found to associated with β1-subunit, although there was a much higher yield of the α1-isoform than of the α2 or α3 (Cone *et al.*, 1991). These experiments suggested that preferences might exist in subunit assembly but that no exclusive isoform selectivity in assembly was found.

More recently we have been studying the assembly of the α-subunit isoforms with another β-subunit, the avian β2-isoform (Lemas *et al.*, 1991). Preliminary experiments that parallel the previous set suggest that β2-subunit can assemble with various α-subunit isoforms but the formation of α1–β2 complexes was very inefficient compared to the formation of α1–β1 complexes. Current experiments include setting up competitions between β1- and β2-subunit isoforms for assembly with each of the α-subunit isoforms.

X. CONCLUSIONS

This chapter presents a summary of ideas and experiments from our lab on the problem of subunit assembly of the Na,K-ATPase and its relation to upregulation. Some aspects of assembly, particulary (1) the possible role of the β-subunit in the proper folding of the α-subunit and (2) intermediate conformations during the assembly process are most elegantly represented in the work of Geering and co-workers (1987, Geering

1990). Experiment on the fates of subunit expressed in *Xenopus* oocytes by Noguchi, Kawamura, and colleagues (Noguchi *et al.*, 1987, 1990) complement the work presented here.

Subunit assembly is an essential step in the biosynthesis of Na,K-ATPase molecules. The assembly process is fast and occurs early in the biosynthetic pathway, before exit of $\alpha-\beta$ complexes from the endoplasmic reticulum. The efficiency of assembly between newly synthesized α- and β-subunit polypeptide chains is not known. Likewise, whether assembly is rate limited by the amounts of just one of the unassembled subunits or by both is not known. Surprisingly, only the parts of the subunits external to the plasma membrane lipid bilayer are currently implicated in direct intersubunit interactions essential for assembly. The regions of the α- and β-subunits' ectodomains involved in subunit interactions are just now being defined. In the α-subunit, some portions of the C-terminal one-third appear to be involved in subunit interactions, but it is not yet know just what portions of this region face outward from the cell surface. There is the possibility that the α-subunit site for interaction with β-subunit varies in structure from one α-subunit isoform to the next. Thus, there may be selectivity in the particular $\alpha-\beta$ complexes that form in cells expressing multiple isoforms of the subunits.

References

Bole, D. G., Hendershot, L. M. and Kearney, J. F. (1986). Posttranslational association of immunoglobulin heavy chain binding protein with nascent heavy chains in nonsecreting and secreting hybridomas. *J. Cell Biol.* **102**, 1558–1566.

Bonifacino, J. S., Cosson, P., and Klausner, R. D. (1990a). Co-localized transmembrane determinants for ER degradation and subunit assembly explain the intracellular fate of RCR chains. *Cell* **63**, 503–513.

Bonifacino, J. S., Suzuki, C. K., and Klausner, R. D. (1990b). A peptide sequence confers retention and rapid degradation in the endoplasmic reticulum. *Science* **247**, 79–82.

Campbell, A. M., Kessler, P. D., and Fambrough, D. M. (1992). The alternative carboxyl termini of avian cardiac and brain SR/ER Ca^{2+}-ATPases are on opposite sides of the membrane. *J. Biol. Chem.*, **267**, 9321–9325.

Clarke, D. M., Loo, T. W., and MacLennan, D. H. (1990). The epitope for monoclonal antibody A20 (amino acids 870–890) is located on the lumenal surface of the Ca^{2+}-ATPase of sarcoplasmic reticulum. *J. Biol. Chem.* **265**, 17405–17408.

Clarke, D. M., Loo, T. W., Inesi, G., and MacLennan, D. H. (1989). Location of the high affinity Ca^{2+}-binding sites within the predicted transmembrane domain of the sacroplasmic reticulum Ca^{2+}-ATPase. *Nature (London)* **339**, 476–478.

Cone, B. C., Takeyasu, K., and Fambrough, D. M. (1991). Structure–function studies of Na/K-ATPase isozymes. *In* "The Sodium Pump: Recent Developments" (P. De Weer and J. Kaplan, eds.), pp. 1265–1269. Rockefeller Univ. Press, New York.

Fambrough, D. M. (1983). Studies on the $(Na^+ + K^+)$-ATPase of skeletal muscle and nerve. *Cold Spring Harbor Symp. Quant. Biol.* **48**, 297–304.

Fambrough, D. M., and Bayne, E. K. (1983). Multiple forms of $(Na^+ + K^+)$-ATPase in the chicken. *J. Biol. Chem.* **258**, 3926–3935.

Fambrough, D. M., Takeyasu, K., Lippincott-Schwartz, J., and Siegel, N. R. (1988). Structure of LEP100, a glycoprotein that shuttles between lysosomes and the plasma membrane, deduced from the nucleotide sequence of the encoding cDNA. *J. Cell Biol.* **106,** 61–67.

Geering, K. (1990). Subunit assembly and functional maturation of Na,K-ATPase. *J. Membr. Biol.* **115,** 109–121.

Geering, K., Kraehenbuhl, J.-P., and Rossier, B. C. (1987). Maturation of the catalytic alpha subunit of Na,K-ATPase during interorganellar transport. *J. Cell. Biol.* **105,** 2613–2619.

Ghosh, S., Freitag, A. C., Martin-Vasallo, P., and Coca-Prados, M. (1990). Cellular distribution and differential gene expression of the three α subunit isoforms of the Na,K-ATPase in the ocular cilary epithelium. *J. Biol. Chem.* **265,** 2935–2940.

Gloor, S., Antonicek, H., Sweadner, K. J., Pagliusi, S., Frank, R., Moos, M., and Schachner, M. (1990). The adhesion molecule on glia (AMOG) is a homologue of the β subunit of the Na,K-ATPase. *J. Cell Biol.* **110,** 165–174.

Jewell, E. A., and Lingrel, J. B. (1991). Comparison of the substrate dependence properties of the rat Na,K-ATPase α1, α2 and α3 isoforms expressed in HeLa cells. *J. Biol. Chem.* **266,** 16925–16930.

Jorgensen, P. L., and Andersen, J. P. (1988). Structural basis for E1-E2 conformational transitions in the Na,K-pump and Ca-pump proteins. *J. Membr. Biol.* **103,** 95–120.

Karin, N. J., Kaprielian, Z., and Fambrough, D. M. (1989). Expression of avian Ca^{2+}-ATPase in cultured mouse myogenic cells. *Mol. Cell. Biol.* **9,** 1978–1986.

Kawakami, K., and Nagano, K. (1988). The transmembrane segment of the Na,K-ATPase β subunit acts as the membrane incorporation signal. *J. Biochem. (Tokyo)* **103,** 54–60.

Kirley, T. (1989). Determination of three disulfide bonds and one free sulfhydryl in the β subunit of (Na,K)-ATPase. *J. Biol. Chem.* **264,** 7185–7192.

Lamb, J. F., and McCall, D. (1972). Effect of prolonged ouabain treatment on Na, K, Cl, and Ca concentration and fluxes in cultured human cells. *J. Physiol. (London)* **225,** 599–617.

Lemas, V., Rome, J., Taormino, J., Takeyasu, K., and Fambrough, D. M. (1991). Analysis of isoform-specific regions within the α- and β-subunits of the avian Na,K-ATPase. *In* "The Sodium Pump: Recent Developments" (P. De Weer and J. Kaplan, eds.), pp. 117–123. Rockefeller Univ. Press, New York.

Lescale-Matys, L., Hensley, C. B., Crnkovic-Markovic, R., Putnam, D. S., and McDonough, A. A. (1990). Low K^+ increases Na,K-ATPase abundance in LLC-PK1/Cl_4 cells by differentially increasing β, and not α, subunit mRNA. *J. Biol. Chem.* **265,** 17935–17940.

Luckie, D. B., Boyd, K. L., and Takeyasu, K. (1991). Ouabain- and Ca^{2+}-sensitive ATPase activity of chimeric Na- and Ca-pump molecules. *FEBS Lett.* **281,** 231–234.

Luckie, D. B., Lemas, V., Boyd, K. L., Fambrough, D. M., Inesi, G., and Takeyasu, K. (1992). Molecular dissection of functional domains of the E1E2-ATPase using sodium and calcium pump chimeric molecules. *Biophys. J.,* **62,** 220–227.

Martin-Vasallo, P., Dackowski, W., Emanuel, J. R., and Levenson, R. J. (1989). Identification of a putative isoform of the Na,K-ATPase β-subunit. Primary structure and tissue-specific expression. *J. Biol. Chem.* **264,** 4613–4618.

Matthews, I., Sharma, R. P., Lee, A. G., and East, J. M. (1990). Transmembranous organization of $(Ca^{2+}-Mg^{2+})$-ATPase from sarcoplasmic reticulum: Evidence for lumenal location of residues 877-888. *J. Biol. Chem.* **265,** 18737–18740.

McDonough, A. A., Geering, K., and Farley, R. A. (1990). The sodium pump needs its β subunit. *FASEB J.* **4,** 1598–1605.

McGrail, K. M., Phillips, J. M., and Sweadner, K. J. (1991). Immunofluorescent localization of three Na,K-ATPase isozymes in the rat central nervous system: Both neurons and glia can express more than one Na,K-ATPase. *J. Neurosci.* 11, 381–391.

Morrow, J. S., Cianci, C. D., Ardito, T., Mann, A. S., and Kashgarian, M. (1989). Ankyrin links fodrin to the alpha subunit of Na,K-ATPase in Madin-Darby canine kidney cells and in intact renal tubule cells. *J. Cell Biol.* 108, 455–465.

Nelson, W. J., and Hammerton, R. W. (1989). A membrane-cytoskeletal complex containing Na^+,K^+-ATPase, ankyrin, and fodrin in Madin–Darby canine kidney (MDCK) cells, implications for the biogenesis of epithelial cell polarity. *J. Cell Biol.* 108, 893–902.

Noguchi, S., Mishima, M., Kawamura, M., and Numa, S. (1987). Expression of functional $(Na^+ + K^+)$-ATPase from cloned cDNAs. *FEBS Lett.* 225, 27–32.

Noguchi, S., Higashi, K., and Kawamura, M. (1990). A possible role of the β-subunit of (Na,K)-ATPase in facilitating correct assembly of the α-subunit into the membrane. *J. Biol. Chem.* 265, 15991–15995.

Pollack, L. R., Tate, E. H., and Cook, J. S. (1981). Na^+,K^+-ATPase in HeLa cells after prolonged growth in low K^+ or ouabain. *J. Cell. Physiol.* 106, 85–97.

Pressley, T. A. (1988) Ion concentration-dependent regulation of Na,K-pump abundance. *J. Membr. Biol.* 105, 187–195.

Price, E. M., and Lingrel, J. B. (1988). Structure-function relationships in the Na,K-ATPase α subunit: Site-directed mutagenesis of glutamine-111 to arginine and asparagine-122 to aspartic acid generates a ouabain-resistant enzyme. *Biochemistry* 27, 8400–8408.

Renaud, K. J., Inman, E. M., and Fambrough, D. M. (1991). Cytoplasmic and transmembrane deletions of Na,K-ATPase β-subunit: Effects on subunit assembly and intracellular transport. *J. Biol. Chem.* 266, 20491–20497.

Rossier, B. C., Geering, K., and Kraehenbuhl, J. P. (1987). Regulation of the sodium pump: How and why? *Trends Biochem. Sci.* 12, 483–487.

Rotundo, R. L. (1988). Biogenesis of acetylcholinesterase molecular forms in muscle. *J. Biol. Chem.* 263, 19398–19406.

Sagara, Y., and Inesi, G. (1991). Inhibition of the sarcoplasmic reticulum Ca^{2+} transport ATPase by thapsigargin at subnanomolar concentrations. *J. Biol. Chem.* 266, 13503–13506.

Shull, G. E., Greeb, J., and Lingrel, J. B. (1986a). Molecular cloning of three distinct forms of the Na^+,K^+-ATPase α-subunit from rat brain. *Biochemistry* 25, 8125–8132.

Shull, G. E., Lane, L. K., and Lingrel, J. B. (1986b). Amino-acid sequence of the β-subunit of the $(Na^+ + K^+)$-ATPase deduced from a cDNA. *Nature (London)* 321, 429–431.

Sweadner, K. J. (1989). Isozymes of the Na^+/K^+-ATPase. *Biochim. Biophys. Acta* 988, 185–220.

Takeda, K., Noguchi, S., Sugino, A., and Kawamura, M. (1988). Functional activity of oligosaccharide-deficient (Na,K)ATPase expressed in *Xenopus* oocytes. *FEBS Lett.* 238, 201–204.

Takeyasu, K., Tamkun, M. M., Siegel, N. R., and Fambrough, D. M. (1987). Expression of hybrid $(Na^+ + K^+)$-ATPase molecules after transfection of mouse Ltk^- cells with DNA encoding the β-subunit of an avian brain sodium pump. *J. Biol. Chem.* 262, 10733–10740.

Takeyasu, K., Tamkun, M. M., Renaud, K. J., and Fambrough, D. M. (1988). Ouabain-sensitive $(Na^+ + K^+)$-ATPase activity expressed in mouse Ltk cells by transfection with DNA encoding the α-subunit of an avian sodium pump. *J. Biol. Chem.* 263, 4347–4354.

Takeyasu, K., Renaud, K. J., Taormino, J. P., Wolitzky, B. A., Barnstein, A., Tamkun, M. M., and Fambrough, D. M. (1989). Differential subunit and isoform expression are involved in regulation of the sodium pump in skeletal muscle. *Curr. Top. Membr. Transp.* 34, 143–165.

Takeyasu, K., Lemas, M. V., and Fambrough, D. M. (1990). Stability of the $(Na^+ + K^+)$-ATPase α-subunit isoforms in evolution. *Am. J. Physiol.* **259,** C619–C630.

Tamkun, M. M., and Fambrough, D. M. (1986). The $(Na^+ + K^+)$-ATPase of chick sensory neurons, studies on biosynthesis and intracellular transport. *J. Biol. Chem.* **261,** 1009–1019.

Taormino, J. P., and Fambrough, D. M. (1990). Pre-translational regulation of the $(Na^+ + K^+)$-ATPase in response to demand for ion transport in cultured chicken skeletal muscle. *J. Biol. Chem.* **265,** 4116–4123.

Wolitzky, B. A., and Fambrough, D. M. (1986). Regulation of the $(Na^+ + K^+)$-ATPase in cultured chick skeletal muscle, modulation of expression by demand for ion transport. *J. Biol. Chem.* **261,** 9990–9999.

CHAPTER 4

Structure–Function Relationship of Na,K-ATPase: The Digitalis Receptor

Cecilia Canessa, Frédéric Jaisser, Jean-Daniel Horisberger, and Bernard C. Rossier
Institut de Pharmacologie et de Toxicologie de l'Université, CH1005 Lausanne, Switzerland

I. INTRODUCTION

Digitalis and related drugs are used for their beneficial action on the heart. Although the mechanism of action was not known, these drugs have been used in clinical medicine for more than 200 years to treat heart failure and cardiac arrhythmias. The discovery that the sodium pump of erythrocytes was selectively inhibited by ouabain, one of the cardiac glycosides used in experimental medicine, indicated that Na,K-ATPase was a major site of action for the drug (Hoffman and Bigger, 1990; Läuger, 1991). How does an inhibition of Na,K-ATPase lead to increased strength of skeletal muscle contraction? A mechanism was proposed which postulates that cardiac glycosides, by inhibiting Na,K-ATPase in the plasma membrane of heart cells, produce an increase in the intracellular sodium which, in turn, leads to an decrease in calcium efflux mediated by the sodium–calcium exchanger (Hoffman and Bigger, 1990). According to this hypothesis, intracellular free calcium would be *indirectly* controlled by the level of sodium pump activity.

Na,K-ATPase is a plasma membrane enzyme made of an α/β hetero-dimer which is present in almost all animal cells (Sweadner, 1989; Lingrel *et al.*,1990; Horisberger *et al.*, 1991). Because of the ubiquitous distribution of the digitalis receptor, cardiac glycosides could influence not only myocardial function but also a number of cellular processes such as cell excitability in brain, skeletal and smooth muscle cells, or ion transport in epithelial cells. This raises the question of how cardiac glycosides could have a selective action on heart without disturbing the function of all cells. Part of the answer resides in the heterogeneity of the digitalis receptor at the molecular level. It has been postulated that Na,K-ATPase activity can be regulated by circulating endogenous ligands often termed endoouabain (Kelly and Smith, 1989,1992; Schoner, 1992). The digitalis receptor may therefore be important not only for the control of heart function but also for a number of cellular processes in brain, in kidney, and in skeletal and smooth muscle. In this short review, we discuss: (i) the general properties of the ligand, the cardiac glycosides, (ii) the structure and the heterogeneity of the digitalis receptor, i.e., the Na,K-ATPase, and (iii) the structure–function relationship of the drug-binding site.

II. GENERAL PROPERTIES OF CARDIAC GLYCOSIDES

Digitalis and related drugs like digoxin, digitoxin, ouabain, and strophantidin are derived from plants commonly known as foxgloves (*Digitalis purpurea* and *Digitalis lanata*) (Hoffman and Bigger, 1990). Other closely related molecules are from animal origins, such as bufalotoxin which is secreted by skin glands of certain toads. Cardiac glycosides have a common structure which is shown in Fig. 1; they are caracterized by a steroid nucleus (5β, 14β-androstane-3β,14-diol) to which an unsaturated lactone side chain is attached at a carbon in position 17 (C17) of the steroid nucleus forming the *aglycone* (meaning without sugar moiety) or *genin* (Hoffman and Bigger, 1990). The aglycone or genin represents the minimal structural requirement to bind to and inhibit Na,K-ATPase (Yoda *et al.*, 1975). By comparing the potency of different natural and synthetic analogues, it has been demonstrated that the inhibitory activity is dependent of the carbonyl oxygen atom on the lactone ring (Yoda and Yoda, 1974). A tight correlation between potency and changes in the position of the carbonyl oxygen atom has been established suggesting that this part of the molecule participates in the binding with the enzyme (Yoda and Yoda, 1974). However, the

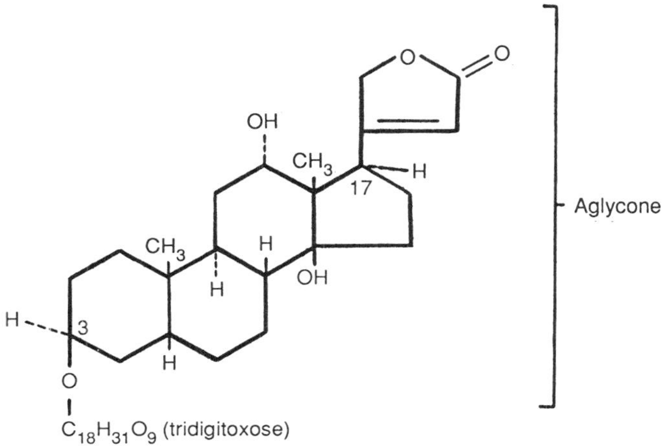

FIGURE 1   Structure of the ouabain molecule.

presence of glycosylated residues (rhamnose, digitoxose, fucose, or cymarose) or a butenolide side chain, attached to the carbon in position 3 (C3) of the steroid nucleus is able to substantially increase inhibitory activity, presumably by allowing the binding of the drug to auxiliary sites on the receptor. The substitution of the 3′-hydroxyl for a 3′-methoxy group in the last sugar residue causes a reduction in stability of the receptor ligand complex (Yoda and Yoda, 1974). Thus, it has been suggested that this 3′-hydroxyl group forms a hydrogen bonding with the enzyme (Yoda and Yoda, 1974). Based on this structure–activity relationship, it has been postulated that the ouabain-binding site on Na,K-ATPase has at least two components: one allows contact of the enzyme with the genin moiety, whereas the other interacts with the sugar moiety and only after a conformational change takes place (Yoda, 1973). When both contact sites are occupied, the drug–enzyme complex becomes stable with half-lives greater than 1 hr. Aglycones lack the sugar moiety and the drug–enzyme complex is less stable with half-lives in the range of minutes (Yoda, 1973).

III. PHYSIOLOGICAL ROLE AND HETEROGENEITY OF THE DIGITALIS RECEPTOR

The digitalis receptor has been amazingly conserved throughout evolution (Horisberger *et al.*, 1991; Canfield *et al.*, 1992; Holzinger *et al.*, 1992; Pressley, 1992). The physiological role of such binding site conservation remains uncertain but several explanations have been proposed. For instance, the digitalis receptor could be directly involved in the catalytic cycle or could play an essential role in pump activity. The existence of highly ouabain-resistant species (rodent,toad) and mutant cell lines with normal pump function does not speak in favor of this hypothesis (Soderberg *et al.*, 1983; Lingrel *et al.*, 1990; Jaisser *et al.*, 1992). A second explanation has been recently proposed. It has been speculated that an endogenous ligand may exist *within* the cell that binds to Na,K-ATPase in intracellular compartments such as endoplasmic reticulum (ER), Golgi apparatus, or endosomes. The occupation of this intracellular binding site would be required for proper subunit oligomerization and transport of Na,K-ATPase from the ER to the plasma membrane (Kelly and Smith, 1989). Until now, no experimental evidence for such mechanism has been presented. A third explanation proposes that the digitalis receptor is an extracellular regulatory binding site for endogenous ligands, controlling enzyme activity (Schoner, 1992; Kelly and Smith, 1992). The existence of endogenous ouabain-like factor(s), or endoouabain in plasma, adrenal gland, and brain, which could play a role in the control of peripheral vascular resistance and blood pressue by preferentially inhibiting the sodium pump in some target tissues, has been recently supported by experimental evidence (Gottlieb *et al.*, 1992; Ferrandi *et al.*, 1992). Thus, the tissue-specific expression of isoforms with different ouabain (or endoouabain) affinities would provide an elegant explanation for the relative selectivity of action of cardiac glycoside (or endoouabain). For instance, the clinical use of cardiac glycosides in the treatment of heart failure could depend on the principle that digitalis binds to a high-affinity receptor in heart so that a low plasma concentration of the drug would be sufficient to bind to the heart receptor, while the low-affinity Na,K-ATPase isoform ubiquitously expressed in all tissues would remain noninhibited (Hoffman and Bigger, 1990). Increasing concentrations of digitalis drugs would induce side effects in the central nervous system, the retina, or the heart by inhibiting progressively low-affinity receptors in these tissues.

The erythrocyte sodium pump or its biochemical counterpart, the Na,K-ATPase, was first identified as the receptor for digitalis (Läuger, 1991). Subsequently, the activity of all Na,K-ATPases from different species and different tissues was found to be inhibited by cardiac glycosides, including digoxin, digitoxin, ouabain, and strophantidin (Yoda *et al.*, 1975; Sweadner, 1989). However, not all Na,K-ATPases display the same affinity for cardiac glycosides. For instance, the rat brain Na,K-ATPase has a more than 100-fold higher sensitivity to ouabain than kidney Na,K-ATPase from the same species (Sweadner, 1989) . Different sensitivities among tissues from the same species have also been observed in many different animals but are generally not as marked as that for rodents (Sweadner, 1989). The observation of different ouabain sensitivity suggests the existence of more than one isoform. The primary structure of the enzyme was obtained by molecular cloning of the Na,K-ATPase α- and β-subunits of sheep (Shul *et al.*, 1985,1986) and fish (Kawakami *et al.*, 1985; Noguchi *et al.*, 1986) . The general structure of Na,K-ATPase α- and β-subunits is shown in Fig. 2. The subsequent cloning and the sequencing of three different Na,K-ATPase α-subunits (α1, α2, α3) in rat and chicken support the idea of functional heterogeneity and experimental evidence for that has been reviewed (Horisberger *et al.*, 1991). An elegant direct demonstration of the differential sensitivity to ouabain of the three rat isoforms has been presented. The expression of recombinant proteins of the rat Na,K-ATPase in baculovirus-infected insect cells allowed the precise measurement of the K_i of ouabain for determined isoforms of the enzyme; α1β1 heterodimers were highly resistant (K_i 91,000 nM), whereas α2β1 and α3β1 were highly sensitive (64 and 23 nM, respectively) (Blanco *et al.*, 1993). Information on the distribution of α-isoforms in different tissues is not yet complete but some rules have been established (Sweadner, 1989). For instance, the α1-subunit is largely distributed and detected in almost all tissues tested. The α2- and α3-isoforms have a much more restricted distribution and they are mainly found in excitable cells, such as heart, skeletal muscle, neurons, and glial cells. The nervous system in general and the brain in particular display the greatest heterogeneity in α isoform expression.

IV. STRUCTURE–FUNCTION RELATIONSHIP OF THE DIGITALIS RECEPTOR

The first approach to the study of structure–function relationship of the digitalis receptor used site-specific ligands, i.e., photoactivable ouabain analogues. These compounds label the part(s) of the protein that make contact with the ligand. The affinity labeling of Na,K-ATPase with various ouabain analogues indicated that the α-subunit is the major component expressing the ouabain-binding site (Forbush, 1983). Several photoactivable derivatives of cardiac glycosides have been synthesized. Photoactivable groups, like NAB (2-nitro-5-azidobenzoil), DAM (ethyl diazomalonyl), NPT (p.nitrophenyltriazene), or NAP (2-nitro-4-azidophenyl) (Goeldner *et al.*, 1983; Rossi *et al.*, 1980; Forbush, 1983), are covalently attached to the genin or sugar moieties of the molecule. NPT-ouabain labels the amino-terminus (a 36-kDa tryptic fragment) of the α polypeptide whereas both 57- and 45-kDa fragments are equally labeled by NAB-ouabain (Rossi

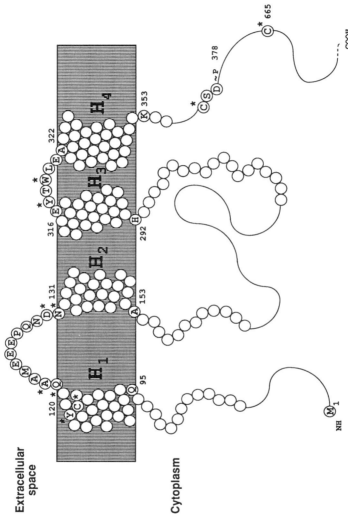

FIGURE 2 Model of the amino-terminal region of the Na,K-ATPase α1-subunit of *Xenopus laevis* containing the putative ouabain-binding site. The hydropathy plot predicts four hydrophobic transmembrane segments (H_1 to H_4) with the N-terminus in the cytoplasmic compartment. The numbering corresponds to the sequence of *X. laevis*. The phosporylation site with the consensus sequence for ion motive P-ATPase (CSDK) is also indicated. *Amino acid residues involved in ouabain binding by a steric or an allosteric mechanism.

et al., 1982; Forbush, 1983). Using a different probe, 24-[³H]azidodigitoxo-side attached to the C17 position, preferential labeling of the second extra-cellular domain (H3–H4 ectocytoplasmic, see Fig. 2) of the α-subunit was detected (McParland *et al.*, 1991). This domain contains a sequence of amino acids well conserved among all species. A small sequence (WLEA) has been tentatively proposed as contact sites (McParland *et al.*, 1991). In all these studies, only the α-subunit is labeled whereas the labeling of the β-subunit appears nonspecific and is detected only when photoreactive groups are attached to the steroid nucleus by long side chains, suggesting that the β-subunit is located somewhat far away from the major binding site on the α-subunit.

The second approach uses the expression of α-subunit from ouabain-resistant species (rat or mouse) into ouabain-sensitive cells. Transfecting α-subunit from a ouabain-resistant species confers ouabain resistance to ouabain-sensitive CV-1 or HeLa cells (Kent *et al.*, 1987; Emanuel *et al.*, 1988; Price and Lingrel, 1988). Conversely, transfection of α-subunit from ouabain-sensitive species to ouabain-resistant recipient cells can confer ouabain sensitivity to the recipient cells (Takeyasu *et al.*, 1988). In con-trast, expressing the β-subunit from a ouabain-resistant species does not confer ouabain resistance (Jaisser *et al.*, 1992). Altogether these results indicate that the α-subunit carries most of the functional domains confer-ring ouabain resistance or sensitivity and confirm the data obtained by the first approach.

The functional domains involved were further defined and identified by engineering chimeric α-subunits. When the amino-terminal parts of an ouabain-resistant species (i.e., rat) were coupled to the carboxy-terminus α-subunit of a ouabain-sensitive species (torpedo, sheep), the resultant chimeric molecule carried the ouabain-resistant phenotype (Price and Lin-grel, 1988; Noguchi *et al.*, 1988). The data stress the critical importance of the amino-terminus of the α-subunit that includes the first four trans-membrane segments (H1–H4) in drug-binding. In order to identify specific amino acids involved in ouabain binding, three strategies have been fol-lowed. First, the comparison of α-subunit sequences from sensitive species (human, sheep, frog) with those from ouabain-resistant species (for in-stance rat, toad) has revealed that the ouabain-resistant form carries charged amino acids at the membrane boundaries of the first extracellular domain (H1–H2 ectodomain; Lingrel *et al.*, 1990; Horisberger *et al.*, 1991; Jaisser *et al.*, 1992). To prove the relevance of these differences, a glutamine (Q111, sheep sequence) and a asparagine (N121) of the ouabain-sensitive sheep α1-subunit were mutagenized for an arginine (Q111R) and an aspartic acid (N121D), the two charged amino acids, characteristic of the ouabain-resistant rat α1-subunit (Price and Lingrel, 1988). The double-

TABLE I

Ouabain-sensitive forms

```
                   ==H1=========|             |=========H2===
α1 Sheep           GAVLCFLAYGI|QAATEEEPQNDN|LYLGVVL
α1 Sheep mut.      •••••••••••|A•••••••••A|•••••••
α1 Human           ••I•••••S•|•••••••••••|•••••••
α1 Pig             ••I•••••••|•••••••••••|•••••••
α1 Chicken         •SL•••••••|TSVM•G••NS••|•••••••
α1 Xenopus         ••I•••••••|•••M•••••••|•••••••
α1 Xenopus mut.    ••I•S•••••|•••M•••••••|•••••••
α1 Torpedo         ••I•••••••|•V••VDN•A•••|•••••••

α2 Human           GAILCFLAYGI|QAAMEDEPSNDN|LYLGVVL
α2 Rat             ••L•••••••|L••••••••••|••••I••
α2 Chicken         •••••••••••|•••••••••••|•••••••
α3 Rat             •••••••••••|••GT•••••G••|••••I••
α3 Chicken         •••••••••••|••GT••••••••|••••I••

α  Drosophila      GAILCFVAYSI|QASTSEEPADDN|LYLGIVL
α  M. sexta                   |••••V•••••• |
α  C. transiens               |••••V•••••• |
```

Ouabain-resistant forms

K_I [μM]

```
                   ==H1=========|             |=========H2===
α1 Rat             GAILCFLAYGI|RSATEEEPPNDD|LYLGVVL     100
α1 Bufo            •••••••••••|RK•SD•••D••N|•••••••      50

α  Hydra           GAILCFFAFGI|RAVRDTNPNMDE|LYLGIVL     100
α  Artemia         •S••••I•YTM|EKYKNPDVLG•N|••••A•        ?
α  D. plexipus                |QASTVEEPSDDH|           >1000

α1 Sheep w.t.      GAVLCFLAYGI|QAATEEEPQNDN|LYLGVVL    <0.1
α1 Sheep mut.      •••••••••••|R•••••••••D|•••••••      100
α1 Sheep mut.      •••••••••••|R••••••••••|•••••••        1
α1 Sheep mut.      •••••••••••|•••••••••D|•••••••         1
α1 Sheep mut.      •••••••••••|D•••••••••R|•••••••      4000
α1 Sheep mut.      •••••••••••|K•••••••••E|•••••••         5
α1 Sheep mut.      •••••••••••|K•••••••••K|•••••••       400
α1 Sheep mut.      •••••••••••|••••••••N•|•••••••          5
α1 Sheep mut.      •••••••••••|••••••••E•|•••••••          5
α1 Sheep mut.      •••••••••••|••••••••K•|•••••••          ?
α1 Sheep mut.      •••••••••••|••••••••A•|•••••••          5
α1 Sheep mut.      •••••••••••|••••••••S•|•••••••          5

α1 Xenopus w.t.    GAILCFLAYGI|QAAMEEEPQNDN|LYLGVVL    <0.1
α1 Xenopus mut.    ••••Y•••••|•••••••••••|•••••••       200
α1 Xenopus mut.    ••••F•••••|•••••••••••|•••••••        80
α1 Xenopus mut.    •••••••••••|R••••••••••|•••••••         1
α1 Xenopus mut.    •••••••••••|RK•••••••••|•••••••         3
α1 Xenopus mut.    •••••••••••|D••••••••••|•••••••      1000
α1 Xenopus mut.    •••••••••••|D•••••••••R|•••••••     >1000
```

Sequences of the outer third of the H1 and H2 putative transmembrane segment and the extracellular loop connecting these two segments. Dots indicate the presence of an amino acid identical to that in the first sequence of the group. Natural isoforms mutant sheep α1-subunit has now a high resistance to ouabain, strongly suggesting that a positive charge at the amino-terminus and negatively charged amino acids at the carboxy-terminus are important for cardiac glycoside-binding function. A summary of the effect of charged amino acids at the border of H1–H2 ectodomain on ouabain affinity of the sheep α1-subunit is presented in Table I. Using the same strategy, we have identified a second positively charged amino acid (a lysine) involved in ouabain binding (Jaisser *et al.*, 1992). Indeed, the sequence of the α1-subunit of the toad *Bufo marinus* is characterized by the presence of two positively charged amino acids at the amino-terminus of the H1–H2 extracytoplasmic domain, and this confers a moderately ouabain-resistant phenotype (Table I).

The second approach consists in site-directed mutagenesis of amino acid residues which are thought to be significant in determining the contact site with a ligand, for instance, the amino acids with hydrogen-bounding potential (Schultheis and Lingrel, 1993). The difficulty of this approach is that a large number of random mutations must be introduced. Many of them will turn out to be either silent or nonfunctional. Using this approach, a aromatic amino acid in the first transmembrane domain (Tyr-108 sheep sequence) was shown to be a determinant of ouabain affinity (Schultheis and Lingrel, 1993).

The third approach takes advantage of the fact that cardiac glycosides can be used to positively select cells which have been mutated in their ouabain-binding site to confer ouabain resistance. Many cell lines with such a phenotype have been described (Soderberg *et al.*, 1983; Cole *et al.*, 1991). We have used the Madin–Darby canine kidney (MDCK) cell line, because of its low, medium, and high resistance to ouabain (Soderberg *et al.*, 1983). A four-step protocol has been followed: first, the cells were exposed to a mutagen (EMS) which introduces randomly point mutations into the genome. Second, ouabain-resistant cells were selected in presence of high concentrations of the drug (up to 5 mM). Third, the α-subunit of the mutated cell line was cloned and sequenced, and relevant mutations were detected. Fourth, the mutations were introduced into a ouabain-sensitive α-subunit (i.e., *X. laevis*) and expressed functionally in a cellular sytem. This approach has the great advantage of allowing only functional

are indicated with normal characters, and mutants generated by site-directed mutagenesis are shown in italic. The ouabain-sensitive forms (with a ouabain K_i lower than 0.1 μM) are in the upper half of the table. The ouabain-resistant forms are in the lower half of the table. Approximate K_i for ouabain are indicated on the right. References for most wild-type sequences can be found in Horisberger *et al.* (1991) and for mutant sequences in: Lingrel *et al.* (1990) Holzinger *et al.* (1992); Jaisser *et al.* (1992); Canfield *et al.* (1992); Baxter-Lowe *et al.* (1989); Canessa *et al.* (1992); C. M. Canessa and B. C. Rossier (unpublished observations).

mutants to be selected. The functional assay allows measurements of the binding parameters of wild-type and mutated subunits (Canessa *et al.*, 1992). Using this approach, we have identified a cysteine in the first transmembrane segment of the α-subunit which confers ouabain resistance (Canessa *et al.*, 1992). When the mutated α1-subunit (C113Y or C113F, *Xenopus* sequence) was introduced in the *X. laevis* α1-subunit of the Na,K-ATPase and expressed in *Xenopus* oocyte, the inhibition constant of ouabain increased more than a 1000-fold, compared with wild type (Table I). A more conservative mutation (C113S) did not change ouabain affinity or pump function, suggesting that this cysteine residue was not involved in a disulfide bond. More recently, we have identified a second amino acid on the H3–H4 ectocytoplasmic loop which has a synergistic effect with the mutation previously observed (C113F) in the first trans-membrane segment (Canessa *et al.*, 1993) (Table II). We have also observed that a conservative change (Y317F) was enough to confer ouabain resistance, again synergistically with the cysteine mutation in the first transmembrane segment (C113Y) (Canessa *et al.*, 1993). We concluded that the removal of a hydroxyl group from the tyrosine was enough to alter significantly ouabain sensitivity.

As summarized in Fig. 2, so far 10 residues have been positively identified as being part of the ouabain receptor: 2 in H1 (C113 and Y117), 4 in the H1–H2 ectocytoplasmic loop (Q120, A121, D130, N131), 2 in the second H3–H4 ectocytoplasmic loop (Y317, W319) and 2 in the second cytoplasmic loop (C376, C665). One should stress that is is still difficult to decide whether the amino acids described here are part of the ouabain-binding site itself, making direct contact with the ligand by steric interaction, or are indirectly involved in ouabain receptor function, i.e., in conformational changes altering the apparent affinity for the drug by an allosteric interaction. Ouabain probably binds to the enzyme in the E2 conformation with two sodium ions occluded within the molecule (Stürmer and Apell, 1992). A mutation could modify the equilibrium between the E1–E2 conformation, thereby causing a change in the apparent affinity for the ligand. Likewise, the dissociation of cardiac glycoside may involve conformational change in the enzyme (Forbush, 1983; Läuger, 1991). Again a mutation modifying this transition could alter the apparent K_i for ouabain. One may speculate, however, that the amino acid residues identified in the H1–H2 and H3–H4 ectocytoplasmic loops are located in a protein domain facing the exterior and in direct contact with the drug which acts from the extracellular side of the membrane. As far as the cysteine in H1 is concerned, we suggested that at least part of the ligand-binding site was embedded in the first membrane helix of the α-subunit, a view supported by

TABLE II

	=H3====|	|====H4=	
α1 mammals, birds	LIL|EYTWLEA|VIF		*
α1 Bufo	•••|H•••••••|•••		**
α1 Xenopus	•••|Q•••••••|•••		
α2 mammals	•••|G•S••••|•••		
α2 Rat,Chicken	•••|G•••••••|•••		
α3 Chicken, α1 Torpedo	•••|G•••••••|•••		
α1 Sheep w.t.	LIL|EYTWLEA|VIF		
α1 Sheep mut.	•••|•••−•••|•••		
α1 Sheep mut.	•••|•••F•••|•••		
α1 Sheep mut.	•••|G•••••••|•••		
α1 Sheep mut.	•••|G••F•••|•••		
α1 Sheep mut.	•••|•••P•••|•••		
α1 Sheep mut.	•••|••C••••|•••		
α1 Sheep mut.	•••|•••••Q•|•••		
α1 Xenopus w.t.	LIL|QYTWLEA|VIF		
α1 Xenopus mut.	•••|•F•••••|•••		***
α1 Xenopus mut.	•••|•C•••••|•••		****

Sequences of the extracellular loop connecting the H3 and H4 putative transmembrane segments. Dots indicate the presence of an amino acid identical to that in the first sequence of the group. Natural isoforms are printed in normal characters and mutants generated by site-directed mutagenesis are printed in italic. References for wild-type sequences can be found in Horisberger et al. (1991) and mutant sequences in: Jaisser et al. (1992); Lingrel et al. (1991); Canessa et al. (1993).

* Includes the ouabain-resistant rat α1-isoform (K_i, 100 μM).

** Ouabain-resistant (K_i, 50 μM).

*** Induces a five-fold increase in the strophanthidin dissociation rate constant.

**** Induces a moderate ouabain resistance by itself, but a very high ouabain resistance when associated with the C113Y mutation in the Xenopus α1-subunit.

recent studies using a monoclonal antibody binding from the extracellular surface (Arystarkhova et al., 1992). On the other hand, it is obvious that two cysteine residues identified in the large cytoplasmic loop (Kirley and Peng, 1991) are not likely to be in contact with the ligand. The segment following H4 has been involved in the coupling between ATP hydrolysis and ion translocation activity in the sarcoplasmic Ca-ATPase (Mac Lennan and Toyofuku, 1992). A mutation of the cysteine close to the phosporylation site (C376, Xenopus sequence, Fig. 2) may impede the important conformational changes which occur during E1–E2 transition.

V. CONCLUSIONS AND PERSPECTIVES

Within a rather short time, a better picture of the ouabain-binding site has emerged. The sequence comparison between ouabain-sensitive and ouabian-resistant species will continue to help define important residues confering ouabain resistance. From this point of view, another charged amino acid in the H1–H2 ectodomain of *Hydra vulgaris* α-subunit was identified (Table I). It could be responsible for the specific pharmacological profile of this species and which is caracterized by a higher sensitivity to strophanthidin than to ouabain (Canfield *et al.*, 1992). The partial cloning of the H1–H2 region of *Danaus plexipus*, a butterfly known to feed on plants with toxic levels of cardiac glycoside and to express a highly ouabain-resistant Na,K-ATPase, is also interesting (Holzinger *et al.*, 1992). An histidine (H131) is an obvious candidate in confering strophanthidin resistance. However, this mutation must be tested in expression systems such as the oocyte or mammalian cells. The selection of a ouabain-resistant cell line could also bring about new information and a number of potential candidates have been identified but not yet functionally tested (Cantley *et al.*, 1992). In our laboratory, we have undertaken a selection of mutants in the presence of different ouabain analogues hoping to select novel mutants with point mutations of residues making contact with specific drug side chains (for instance the sugar moieties) which are the caracteristic of a variety of cardiac glycosides, altering their binding properties (Yoda and Yoda, 1974). Ultimately, the 3D structure of the ion-motive P-ATPases will be solved and the information gathered by the indirect strategies described in this review will become more important.

References

Arystarkhova, E., Gasparian, M., Modyanov, N. N., and Sweadner, K. J. (1992). Na,K-ATPase extracellular surface probed with a monoclonal antibody that enhances ouabain binding. *J. Biol. Chem.* **267**, 13694–13701.

Baxter-Lowe, L. A., Guo, J. Z., Bergstrom, E. E., and Hokin, L. E. (1989). Molecular cloning of the Na,K-ATPase α-subunit in developing brine shrimp and sequence comparison with higher organisms. *FEBS Lett.* **257**, 181--87.

Blanco, G., Jian Xie, J., and Mercer, R. W. (1993). Functional expression of the $\alpha2$ and $\alpha3$ isoforms of the Na, K-ATPase in baculovirus-infected insect cells. *Proc. Natl. Acad. Sci. U.S.A.* **90**, 1824–1828.

Canessa, C. M., Horisberger, J. D., Louvard, D., and Rossier, B. C. (1992). Mutation of a cysteine in the first transmembrane segment of Na, K-ATPase alpha subunit confers ouabain resistance. *EMBO J.* **11**, 1681–1687.

Canessa, C. M., Horisberger, J.-D., and Rossier, B. C. (1993). Mutation of a tyrosine in the H3-H4 ectodomain of Na, K-ATPase α subunit confers ouabain resistance. *J. Biol. Chem.*, **268**, 17722–17726.

Canfield, V. A., Xu, K., D'Aquilla, T., Shyjan, A. W., and Levenson, R. (1992). Molecular cloning and caracterization of Na, K-ATPase from *Hydra vulgaris:* Implications for enzyme evolution and ouabain sensitivity. *New Biol.* **4,** 339–348.

Cantley, L. G., Zhou, X. M., Cunha, M. J., Epstein, J., and Cantley, L. C. (1992). Ouabain-resistant transfectants of the murine ouabain resistance gene contain mutations in the alpha-subunit of the Na,K-ATPase. *J. Biol. Chem.* **267,** 17271–17278.

Cole, J., Richmond, F. N., and Bridges, B. A. (1991). The mutagenicity of 2-amino-N^6-hydroxyadenine to L5178Y $tk^{+/-}$ 3.7.2C mouse lymphoma cells: Measurement of mutations to ouabain, 6-thioguanine and trifluorothymidine resistance, and the induction of micronuclei. *Mutat. Res.* **253,** 55–62.

Emanuel, J. K., Schulz, J., Zhou, X. M., Kent, R., Housman, D., Cantley, L., and Levenson, R. (1988). Expression of an ouabain-resistant Na, K-ATPase in CV-1 cells after transfection with a cDNA encoding the rat Na, K-ATPase α1 subunit. *J. Biol. Chem.* **263,** 7726–7733.

Ferrandi, M., Minotti, E., Salardi, S., Florio, M., Bianchi, G., and Ferrari, P. (1992). Ouabainlike factor in Milan hypertensive rats. *Am. J. Physiol.* **263,** F739–F748.

Forbush, B.I.I.I. (1983). Cardiotonic steroid binding to Na, K-ATPase. *Curr. Top. Membr. Transp.* **19,** 167–201.

Goeldner, M. P., Hirth, C. G., Rossi, B., Ponzio, G., and Lazdunski, M. (1983). Specific photoaffinity labeling of the digitalis binding site of the sodium and potassium ion activated adenosinetriphosphatase induced by energy transfer. *Biochemistry* **22,** 4685–4690.

Gottlieb, S. S., Rogowski, A. S., Weinberg, M., Krichten, C. M., Hamilton, B. P., and Hamlyn, J. M. (1992). Elevated concentrations of endogenous ouabain in patients with congestive heart failure. *Circulation* **86,** 420–425.

Hoffman, B. F., and Bigger, J. T. (1990). Digitalis and allied cardiac glycosides. *In* "The Pharmacological Basis of Therapeutics" (A. Goodman Gilman, T. W. Rall, A. S. Nies, and P. Taylor, eds.), pp. 814–839. Pergamon, New York.

Holzinger, F., Frick, C., and Wink, M. (1992). Molecular basis for the insensivity of the Monarch (*Danaus plexippus*) to cardiac glycosides. *FEBS Lett.* **314,** 477–480.

Horisberger, J.-D., Lemas, V., Kraehenbuhl, J.-P., and Rossier, B. C. (1991). Structure–function relationship of Na,K-ATPase. *Annu. Rev. Physiol.* **53,** 565–584.

Jaisser, F., Canessa, C. M., Horisberger, J. D., and Rossier, B. C. (1992). Primary sequence and functional expression of a novel ouabain-resistant Na,K-ATPase—The beta-subunit modulates potassium activation of the Na,K-pump. *J. Biol. Chem.* **267,** 16895–16903.

Kawakami, K., Noguchi, S., Noda, M., Takahashi, H., Ohta, T., Kawamura, M., Nojima, H., Nagano, K., Hirose, T., Inayama, S., Hayashida, H., Miyata, T., and Numa, S. (1985). Primary structure of the α-subunit of *Torpedo california* ($Na^+ + K^+$)ATPase deduced from cDNA sequence. *Nature (London)* **316,** 733–736.

Kelly, R. A., and Smith, T. W. (1989). The search for the endogenous digitalis: An alternative hypothesis. *Am. J. Physiol.* **256,** C937–C950.

Kelly, R. A., and Smith, T. W. (1992). Is ouabain the endogenous digitalis. *Circulation* **86,** 694–697.

Kent, R. B., Emanuel, J. R., Neriah, Y. B., Levenson, R., and Housman, D. E. (1987). Ouabain resistance conferred by expression of the cDNA for a murine Na^+, K^+-ATPase α-subunit. *Science* **237,** 901–903.

Kirley, T. L., and Peng, M. (1991). Identification of cysteine residues in lamb kidney (Na,K)-ATPase essential for ouabain binding. *J. Biol. Chem.* **266,** 19953–19957.

Läuger, P. (1991). "Electrogenic Ion Pump." Sinauer, Sunderland, Massachusetts.

Lingrel, J. B., Orlowski, J., Shull, M. M., and Price, E. M. (1990). Molecular genetics of Na, K-ATPase. *Prog. Nucleic Acid Res. Mol. Biol.* **38,** 37–89.

Lingrel, J. B., Orlowski, J., Price, E. M., and Pathack, B. G. (1991). Regulation of the α-subunit genes of the Na, K-ATPase and determinant of cardiac glycoside sensitivity. *In* "The Sodium Pump: Structure, Mechanism, and Regulation" (J. H. Kaplan and P. De Weer, eds.), pp. 1–16. Rockfeller Univ. Press, New York.

MacLennan, D. H., and Toyofuku, T. (1992). Structure–function relationships in the Ca2+ pump of the sarcoplasmic reticulum. *Biochem. Soc. Trans.* **20,** 559–562.

McParland, R. A., Fullerton, D. S., Becker, R. R., From, A. H. L., and Ahmed, K. (1991). Studies of digitalis binding site (s) in Na, K-ATPase by covalent labeling with a photoactive probe located in the genin moiety. *In* "The Sodium Pump: Recent Developments" (J. H. Kaplan and P. De Weer, eds.), pp. 297–302. Rockfeller Univ. Press, New York.

Noguchi, S., Noda, M., Takahashi, H., Numa, S., Kawakami, K., Ohta, T., *et al.* (1986). Primary structure of the b-subunit of Torpedo california (Na+, K+)-ATPase deduced from the cDNA sequence. *FEBS Lett.* **320,** 315–320.

Noguchi, S., Ohta, T., Takeda, K., Ohtsubo, M., and Kawamura, M. (1988). Ouabain sensitivity of a chimeric α subunit (*Torpedo*/rat) of the (Na, K)-ATPase expressed in *Xenopus* oocyte. *Biochem. Biophys. Res. Commun.* **155,** 1237–1243.

Pressley, T. A. (1992). Phylogenetic conservation of isoform-specific regions within α-subunit of Na+, K+-ATPase. *Am. J. Physiol. Cell. Physiol.* **262,** C743–C751.

Price, E. M., and Lingrel, J. B. (1988). Structure–function relationships in the Na, K-ATPase alpha subunit: Site-directed mutagenesis of glutamine-111 to arginine and asparagine-122 to aspartic acid generates a ouabain-resistant enzyme. *Biochemistry* **27,** 8400–8408.

Rossi, B., Vuilleumier, P., Gache, C., Balerna, M., and Lazdunski, M. (1980). Affinity labelling of the digitalis receptor with p-nitrophenytriazene-ouabain, a highly specific alkylating agent. *J. Biol. Chem.* **255,** 9936–9941.

Rossi, B., Ponzio, G., and Lazdunski, M. (1982). Identification of the segment of the catalytic subunit of (Na+, K+)APTase containing the digitalis binding site. *EMBO J.* **1,** 859–861.

Schoner, W. (1992). Endogenous digitalis-like factors. *Clin. Exp. Hypertens.* [A] **14,** 767–814.

Schultheis, P. J., and Lingrel, J. B. (1993). Substitution of transmembrane residues with hydrogen-bonding potential in the α-subunit of Na, K-ATPase reveals alterations in ouabain sensitivity. *Biochemistry* **32,** 544–550.

Shul, G. E., Schwartz, A., and Lingrel, J. B. (1985). Amino-acid sequence of the catalytic subunit of the (Na+ K+) ATPase deduced from a complementary DNA. *Nature (London)* **316,** 691–695.

Shull, G. E., Lane, L. K., and Lingrel, J. B. (1986). Amino-acid sequence of the b-subunit of the (Na+, K+)ATPase deduced from a cDNA. *Nature (London)* **321,** 429–431.

Soderberg, K., Rossi, B., Lazdunski, M., and Louvard, D. (1983). Characterization of ouabain-resistant mutants of a canine kidney cell line, MDCK. *J. Biol. Chem.* **258,** 12300–12307.

Stürmer, W., and Apell, H. J. (1992). Fluorescence study on cardiac glycoside binding to the Na, K pump. Ouabain binding is associated with movement of electrical charge. *FEBS Lett.* **300,** 1–4.

Sweadner, K. J. (1989). Isozymes of the Na+/K+-ATPase. *Biochim. Biophys. Acta* **988,** 185–220.

Takeyasu, K., Tamkun, M. M., Renaud, K. J., and Fambrough, D. M. (1988). Ouabain-sensitive (Na+ + K+)-ATPase activity expressed in mouse L cells by transfection with DNA encoding the α-subunit of an avian sodium pump. *J. Biol. Chem.* **263,** 4347–4354.

Yoda, A. (1973). Structure–activity relationships of cardiotonic steroids for the inhibition of sodium- and potassium-dependent adenosine triphosphatase. I. Dissociation rate constants of various enzyme-cardiac glycoside complexes formed in the presence of magnesium and phosphate. *Mol. Pharmacol.* **9,** 51–60.

Yoda, A., and Yoda, S. (1974). Structure–activity relationships of cardiotonic steroids for the inhibition of sodium- and potassium-dependent adenosine triphosphatase. 3. Dissociation rate constants of various enzyme–cardiac glycoside complexes formed in the presence of sodium, magnesium, and adenosine triphosphate. *Mol. Pharmacol.* **10,** 494-500.

Yoda, S., Sarrif, A. M., and Yoda, A. (1975). Structure–activity relationships of cardiotonic steroids for the inhibition of sodium- and potassium-dependent adenosine triphosphatase. IV. Dissociation rate constants for complexes of the enzyme with cardiac oligodigitoxides. *Mol. Pharmacol.* **11,** 647–652.

PART III

Sorting of Ion Transport Proteins
and the Creation of Polarized
Membrane Domains

CHAPTER 5

Subcellular Targeting and Regulation of Glucose Transporters

Peter M. Haney* and Mike Mueckler†
Departments of *Pediatrics and *,†Cell Biology and Physiology, Washington University
School of Medicine, St. Louis, Missouri 63110

I. INTRODUCTION

A. Cellular Glucose Uptake Is Mediated by a Family of Five Transporter Isoforms

Glucose is an essential source of chemical energy for many cells as well as a precursor for the biosynthesis of NADPH, nucleotides, and complex carbohydrates and glycoproteins. Facilitated diffusion of glucose across membranes is mediated by a family of five glucose transporter isoforms (Bell *et al.*, 1990; Mueckler, 1990). These membrane glycoproteins are named GLUT1 through GLUT5 in the order of their discovery by cDNA cloning. The five isoforms are distinct in their tissue distributions, kinetic properties, and regulation (Devaskar and Mueckler, 1992). Hydropathy plot analysis (Mueckler *et al.*, 1985) suggests that the glucose transporters

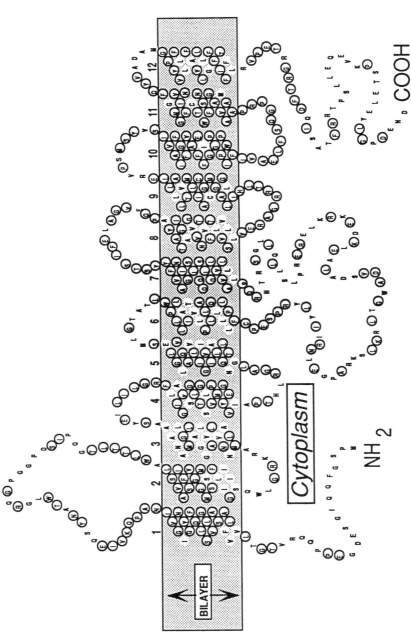

FIGURE 1 Predicted topology of the rat GLUT4 glucose transporter in the plasma membrane. Amino acid residues are given by the single-letter code. Residues that are identical between GLUT1 and GLUT4 are circled.

consist of 12 membrane-spanning α-helical domains connected by hydrophilic segments. Each transporter has a large cytoplasmic loop between the sixth and seventh transmembrane domains. Each also has a large exoplasmic loop between the first and second transmembrane domains that contains an N-glycosylation site. There is a high degree of homology among members of the family. For example, the human GLUT1 and GLUT4 proteins exhibit 65.3% sequence identity with an additional 11.9% sequence similarity representing conservative amino acid differences. Figure 1 shows the predicted topology of the glucose transporters and depicts the homology between GLUT1 and GLUT4.

Glucose transport in many cell types is not a rate-limiting step for overall cellular glucose metabolism and is therefore not subject to acute regulation. For example, the major glucose transporter isoforms expressed in endothelial cells, hepatocytes, and neurons (GLUT1, GLUT2, and GLUT3, respectively) are expressed constitutively at high levels in the plasma membrane. The flux of glucose across the membrane of these cell types varies primarily according to the circulating glucose concentration and is largely unaffected by changes in the levels of hormones and other circulatory factors. On the other hand, the so-called insulin-sensitive tissues, muscle and fat, respond within minutes to elevated blood insulin levels with a dramatic increase in cellular glucose uptake (Gibbs *et al.,* 1988).

Muscle and fat cells express both GLUT1 and GLUT4. The expression of GLUT4 is restricted to insulin-sensitive cell types and this isoform appears to be responsible for most if not all of the insulin-mediated increase in glucose transport (Birnbaum, 1989; Charron *et al.,* 1989; Fukumoto *et al.,* 1989; James *et al.,* 1989). The effect of insulin on glucose transport in fat and muscle is made possible by the intracellular sequestration of GLUT4 in the absence of the hormone. The rapid response of GLUT4 to insulin is critical for the maintenance of normal glucose homeostasis, because skeletal muscle is the major depot for sugar disposal in the postprandial state (Ferrannini *et al.,* 1985; James *et al.,* 1985). Thus, unraveling the subcellular trafficking of glucose transporters in insulin-sensitive cell types is a prerequisite for understanding the mechanism by which insulin regulates blood glucose levels.

B. GLUT4 Is Targeted to Intracellular Compartments in Muscle and Fat

The subcellular distribution of glucose transporters in insulin-sensitive cells has been studied by differential centrifugation of cellular homogenates (Cushman and Wardzala, 1980; Suzuki and Kono, 1980; Hirshman *et al.,* 1990; Piper *et al.,* 1991a), cell surface labeling (Gould *et al.,* 1989; Holman *et al.,* 1990), and morphologic methods such as immunofluorescence (Tordjman *et al.,* 1990) and electron microscopy of immunogold-labeled specimens (Slot *et al.,* 1991b). GLUT1 is present in both intracellular membranes and the plasma membrane of basal adipocytes (Piper *et al.,* 1991a). In contrast, GLUT4 appears to be associated almost exclusively with intracellular membranes in non-insulin-treated adipocytes (Birnbaum, 1989; James *et al.,* 1989; Piper *et al.,* 1991a) For example, in untreated 3T3-L1 adipocytes the ratio of the amount of GLUT4 per unit protein in plasma membranes vs intracellular membranes is 1:30 (Piper *et al.,* 1991a). Similarly, immunogold labeling of frozen ultrathin sections was used to demonstrate that 99% of GLUT4 is located within cytoplasmic structures of unstimulated brown adipose tissue. These latter studies revealed that intracellular GLUT4 is located predominately in tubulovesicular elements of the *trans*-Golgi reticulum (Slot *et al.,* 1991b). Slot and colleagues have presented evidence that the bulk of GLUT4 protein in unstimulated skeletal and cardiac muscle is also present in cytoplasmic tubulovesicular structures (Slot *et al.,* 1990,1991a). Smith *et al.* (1991) reported that 95% of GLUT4 detected by immunogold labeling of basal white adipocytes is present in small intracellular vesicles contiguous with or in close proximity to the plasma membrane. Thus, it is clear that GLUT4 is very efficiently excluded from the plasma membrane in the absence of insulin, but there may be some heterogeneity among cell types in the morphologic appearance of the intracellular GLUT4-containing compartments.

C. Insulin Stimulates Glucose Transport by Causing Translocation of GLUT4

Two groups simultaneously advanced the "translocation" hypothesis to account for the action of insulin on adipocyte glucose transport. Cushman and Wardzala (1980) demonstrated a stoichiometric relationship between a decrease in the glucose-displaceable binding of cytochalasin B (a specific assay for facilitative glucose transporters) to microsomal membranes and an increase in binding to plasma membranes isolated from rat adipocytes treated with insulin. Suzuki and Kono (1980) reported similar findings using reconstitution of glucose transport activity from subcellular membrane fractions. Both groups suggested that the redistribution or translocation of glucose transporters from an intracellular compartment to the plasma membrane plays an important role in the effect of insulin on transport activity. Direct morphologic evidence for the translocation of GLUT4 from its intracellular compartment to the cell surface has recently been obtained. Insulin exposure increases the fraction of cellular

GLUT4 present in the plasma membrane from 1 to 40% in brown adipose tissue (Slot *et al.*, 1991b). Calderhead *et al.* (1990) addressed the significance of transporter translocation for insulin responsiveness by labeling cell surface transporters of basal and insulin-treated 3T3-L1 adipocytes with a specific membrane-impermeant, photoaffinity-labeling reagent. Labeling of GLUT4 at the cell surface increased 17-fold after insulin treatment. This was accompanied by a 21-fold increase in hexose transport. GLUT1 constituted 75% of the total glucose transporter population, and its labeling increased 6.5-fold in response to insulin. The results indicate that the translocation of GLUT4 could largely account for the insulin effect on transport rate, but only if the intrinsic activity of GLUT4 is much higher than that of GLUT1. Clark *et al.* (1991), in a similar study, attributed a 30-fold stimulation of glucose transport by insulin in rat adipocytes primarily to a 15-fold increase in plasma membrane GLUT4. They suggested the remaining discrepancy may be due to an ~2 fold stimulation of the intrinsic activity of GLUT4. There is indirect evidence that the intrinsic activity of glucose transporters may be regulated under certain conditions (Kahn *et al.*, 1988,1991; Clancy and Czech, 1990; Clancy *et al.*, 1991.

Technical limitations have made it difficult to address the dynamics and kinetics of glucose transporter trafficking in insulin-sensitive cells. Clark *et al.* (1991) showed that the $t_{1/2}$ for internalization of cell surface-labeled GLUT4 after insulin withdrawal in rat adipocytes is 9–11 min. Slot *et al.* (1991b) were able to visualize GLUT4 in coated pits, coated vesicles, and early endosomes of insulin-stimulated brown adipose tissue using immunogold labeling and electron microscopy. On the basis on these data, they propose that GLUT4 recycles continuously from its intracellular compartment to the plasma membrane and that insulin elicits the redistribution of the transporter by stimulating exocytosis rather than by slowing the rate of endocytosis. Thus, the trafficking of GLUT4 may, to a certain extent, resemble the well-characterized recycling pathway of many cell surface receptors, such as the low-density lipoprotein (Goldstein *et al.*, 1985) and transferrin (Stoorvogel *et al.*, 1987) receptors. Double immunogold-labeling experiments will help clarify the relationship between the trafficking pathways of the glucose transporters and various membrane receptors.

II. INTRACELLULAR TARGETING OF GLUT4 IS ISOFORM-SPECIFIC AND INDEPENDENT OF CELL TYPE

Insulin causes translocation of both GLUT1 and GLUT4 from intracellular compartments to the cell surface in rat adipocytes and 3T3-L1 adipocytes (Zorzano *et al.*, 1989; Calderhead *et al.*, 1990; Holman *et al.*, 1990; Piper *et al.*, 1991a). Movement of both isoforms suggests that cell-specific rather than isoform-specific factors confer the potential for insulin-regulated translocation. However, there are targeting differences between GLUT1 and GLUT4 in insulin-sensitive cells. The two isoforms are present in distinct populations of intracellular vesicles in adipocytes (Piper *et al.*, 1991a; Zorzano *et al.*, 1989). The intracellular sequestration of GLUT4 is far more efficient than that for GLUT1 in basal cells. Consequently, the relative increase at the cell surface in response to insulin is much greater for GLUT4 than for GLUT1. Thus, GLUT4 seems particularly adapted to acute regulation by rapid changes in its subcellular distribution.

Until recently it was not known whether the capacity of GLUT4 for acute regulation is the consequence of a cell-specific apparatus for targeting and translocation, or, alternatively, reflects the presence of a ubiquitous cellular machinery that acts in an isoform-specific manner. We tested the hypothesis that factors specific to insulin-sensitive tissues are required for glucose transporter targeting by expressing GLUT4 in two cell types that do not normally contain GLUT4 and do not exhibit insulin-dependent changes in glucose transport (Haney *et al.*, 1991). 3T3-L1 cells replicate in culture as fibroblast-like preadipocytes. After achieving confluence, they can be induced to differentiate into insulin-sensitive adipocyte-like cells (Green and Kehinde, 1974). 3T3-L1 preadipocytes express only the GLUT1 isoform and do not respond to insulin with an increase in glucose uptake. During differentiation, GLUT4 appears, the level of GLUT1 decreases, and glucose uptake becomes exquisitely sensitive to insulin (Tordjman *et al.*, 1989). The second cell type we used was the human hepatoma cell line, HepG2. These cells also express only GLUT1 and do not exhibit insulin-stimulated glucose uptake, but they do express insulin receptors and respond to the hormone in other ways (Adeli and Sinkevitch, 1990; Conover and Lee, 1990).

3T3-L1 fibroblasts were transfected with bovine papillomavirus plasmid vectors containing either human GLUT1 or rat GLUT4 cDNAs downstream of the murine metallothionein-I promoter. A plasmid encoding neomycin resistance was cotransfected to provide the ability to select transfectants in G418 (Gorman, 1985). Cell lines overexpressing each transporter, as well as control lines transfected only with the neomycin resistance plasmid, were selected for further study.

Parental 3T3-L1 preadipocytes and control transfected cell lines contained 170 ng GLUT1 per mg total protein. Two 3T3-L1 cell lines overexpressing GLUT1, 3T3-Hep 18 and 3T3-Hep 4, were characterized. The level of GLUT1 was increased 1.6- and 3.2-fold in these lines relative to controls. Western blot analysis using a monoclonal antibody specific for human GLUT1 and not reactive with the equivalent murine isoform showed that the increase in GLUT1 in these lines was due to the expression of the heterologous human protein and not to variations in endogenous GLUT1. Consistent with the differential expression of GLUT1 in these lines, glucose uptake was increased by 2.7- and 4.1-fold, respectively, in 3T3-Hep 18 and 3T3-Hep4 relative to Neo-1, a control transfected cell line (Fig. 2). The increase in glucose transport activity suggests that exogenous GLUT1 is appropriately targeted to the plasma membrane in these cells.

GLUT4 was not detected in parental 3T3-L1 fibroblasts. The cell line designated 3T3-IRGT25 expressed 60 ng of GLUT4 per mg total protein. This is quite similar to the value of 65 ng of GLUT4 per mg total protein in differentiated 3T3-L1 adipocytes, which exhibit a 15-fold stimulation of glucose uptake in response to insulin. However, neither the cell line 3T3-IRGT25 nor a second transfected line expressing less GLUT4 showed any evidence of altered basal glucose uptake or of an insulin response (Fig. 3). Expression of GLUT4 in these cells is driven by the murine metallothionein-I promoter of the bovine papilloma virus vector, and thus treatment with butyrate and zinc induces transcription. 3T3-IRGT25 cells incubated overnight with 7mM sodium butyrate and 100 μM zinc chloride expressed 198 ng of GLUT4 per mg protein. These cells exhibited a >twofold increase in total glucose transporters relative to parental preadipocytes and control transfectants but showed no change in basal glucose uptake or evidence of an insulin response. This strongly suggests that GLUT4 in these cells is not targeted to the plasma membrane in the basal

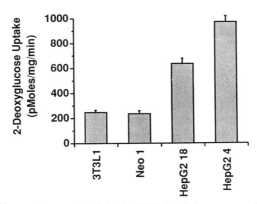

FIGURE 2 Glucose transport activity in 3T3-L1 cell lines expressing human GLUT1. Values for 2-[^{3}H]deoxyglucose uptake represent means ±SEM of six determinations. (Reproduced from the *Journal of Cell Biology*, 1991, vol. 114, p. 692, by copyright permission of the Rockefeller University Press.)

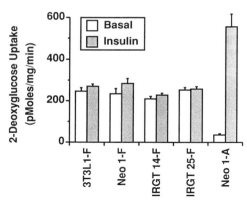

FIGURE 3   Glucose transport activity in 3T3-L1 cell lines expressing rat GLUT4. Cells were either untreated or treated with 100 nM insulin for 20 min. Values for 2-[3H]deoxyglucose uptake represent means ±SEM of six determinations. (Reproduced from the *Journal of Cell Biology*, 1991, Vol. 114, p. 693, by copyright permission of the Rockefeller University Press.)

state or translocated to the cell surface in response to insulin, but rather is sequestered within the cell.

The overexpressing cell lines, 3T3-Hep4 and 3T3-IRGT25, were studied by confocal immunofluorescence microscopy to directly determine the subcellular localization of glucose transporters. The plasma membrane was most prominently labeled in 3T3-Hep4 cells incubated with GLUT1 antibody, but some intracellular staining was also evident (Fig. 4). In sharp contrast, 3T3-IRGT25 cells incubated with GLUT4 antibody showed localization almost entirely within cytoplasmic structures. The label was concentrated in a perinuclear region and also distributed in a punctate pattern throughout the cytoplasm (Fig. 4). This pattern strongly resembles that seen for endogenous GLUT4 in differentiated 3T3-L1 adipocytes (Piper *et al.*, 1991a; Mueckler and Tordjman, 1991). We saw no evidence of GLUT4 translocation in transfected preadipocytes after insulin expo-

FIGURE 4   Confocal immunofluorescent images of GLUT1 (left) and GLUT4 (right) transfected fibroblasts, at lower (bottom) and higher (top) magnifications. 3T3-Hep4 cells incubated with antibody mixed with an excess of a rat COOH-terminal GLUT1 peptide exhibited no immunofluorescence, and the control transfectant Neo-1 line showed no immunofluorescence when incubated with GLUT4 antibody. Bars: top right and bottom, 25 μm; top left, 10 μm. (Reproduced from the *Journal of Cell Biology*, 1991, vol. 114, p. 692, by copyright permission of the Rockefeller University Press.)

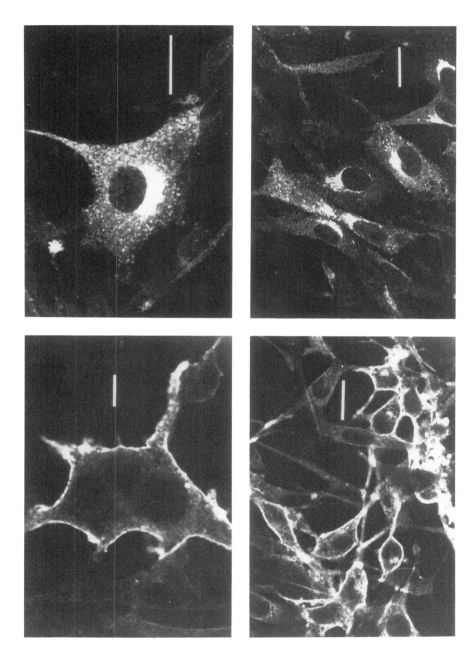

sure. The differential targeting of GLUT1 and GLUT4 in these cells was confirmed by sucrose density gradient analysis (Haney *et al.*, 1991).

The intracellular location of GLUT4 in 3T3-IRGT25 cells was further defined by immunogold labeling and electron microscopy of frozen ultrathin sections. Tubulovesicular elements, often in close proximity to Golgi stacks, were heavily stained, with almost no labeling of the cell surface. This is very similar to the pattern observed for endogenous GLUT4 in a variety of insulin-sensitive cells in the basal state, including brown adipocytes (Slot *et al.*, 1991b) and cardiac myocytes (Slot *et al.*, 1991a).

HepG2 cells were transfected by lipofection with pSSFV-GLUT4, a plasmid which confers neomycin resistance and includes the splenic focus-forming virus LTR (Yamamoto *et al.*, 1981) to drive expression of GLUT4. Immunogold labeling of endogenous GLUT1 clearly demonstrated localization of the transporter to the plasma membrane. Sucrose density gradient analysis revealed distinct patterns of distribution for GLUT1 and GLUT4. GLUT1 was present almost exclusively in the very dense (1.15 g/cm^3) fractions at the bottom of the gradient, corresponding to the plasma membrane. These fractions contained no GLUT4. In basal and insulin-exposed cells, GLUT4 was found in a fraction corresponding to an intracellular compartment of the same density (mean 1.09 g/cm^3) as that seen in basal 3T3-IRGT25 fibroblasts, 3T3 adipocytes, and rat adipocytes. Thus, the targeting of GLUT1 and GLUT4 in HepG2 cells appears to closely resemble their differential targeting in 3T3-L1 preadipocytes.

These experiments were performed to test the hypothesis that factors specific to insulin-sensitive tissues are required for the proper intracellular targeting of GLUT4. The data from transfected 3T3-L1 fibroblasts and HepG2 cells suggest that non-insulin-sensitive cells correctly target GLUT4, rejecting this hypothesis. In support of this conclusion, NIH-3T3 fibroblasts and fibroblast cell lines overexpressing insulin receptors also target GLUT4 properly (Hudson *et al.*, 1992). On the other hand, the results also suggest that factors specific to insulin-sensitive tissues must be required for insulin-dependent translocation of glucose transporters to the cell surface.

III. THE STRUCTURAL REQUIREMENTS FOR GLUT4 TARGETING ARE UNDER INVESTIGATION

GLUT4 likely contains a subcellular targeting motif, absent in other glucose transporter isoforms, that is recognized by a ubiquitously expressed cellular targeting apparatus. The N- and C-terminal cytoplasmic domains of GLUT4 are distinct from those of the other isoforms and are therefore the regions most likely to contain a sorting motif. Two recent reports suggest that transmembrane domains are involved in the retention of proteins in the *trans*-Golgi network (Munro, 1991; Nilsson *et al.*, 1991). However, the high degree of homology among the transmembrane segments of the glucose transporter isoforms may preclude their function as targeting signals.

Current work centers on defining a sorting motif by studying the subcellular targeting of chimeric glucose transporter molecules. A preliminary report (Piper *et al.*, 1991b) suggests that the N-terminal 38 amino acids of GLUT4, that is, the N-terminal cytoplasmic domain as well as most of the first transmembrane segment, are sufficient to confer intracellular targeting on a chimera otherwise consisting of GLUT1. These studies examined the behavior of GLUT4 in transiently transfected insulin-insensitive cell lines. Confirmation of the N-terminus as a signal that is both necessary and sufficient for the targeting of GLUT4 to an insulin-sensitive intracellular compartment will require expression of chimeric membrane proteins containing this putative motif in insulin-sensitive cells in which effects on glucose transport activity, distribution in subcellular fractions, and immunocytochemical localization can be observed in the presence and absence of the hormone. These experiments may also delineate the structural determinants of glucose transporter endocytosis, which may or may not be identical to a *trans*-Golgi sorting signal (see below). Internalization signals of four constitutively recycling receptors—the cation-dependent mannose-6-phosphate receptor (Johnson *et al.*, 1990), the cation-independent mannose-6-phosphate receptor (Canfield *et al.*, 1991), the transferrin receptor (Collawn *et al.*, 1990), and the low-density lipoprotein receptor (Chen *et al.*, 1990)—consist of four to six amino acids that form a "tight turn" within a cytoplasmic tail (Trowbridge, 1991). There are no homologous signals within glucose transporters, but these studies offer a model for the analysis of endocytic targeting information. We speculate that polytopic membrane proteins such as GLUT4 may possess multiple signals that act in concert to determine subcellular distribution under different physiological conditions.

Whether structural elements of GLUT4 are important in its insulin-dependent redistribution to the plasma membrane is not known. The intracellular compartment containing GLUT4 has been characterized in a variety of insulin-sensitive cells, including 3T3-L1 adipocytes (Biber and Lienhard, 1986; Piper *et al.*, 1991a), rat adipocytes (James *et al.*, 1987; Slot *et al.*, 1991b), and rat cardiac myocytes (Slot *et al.*, 1991a). Some of these cells are known to contain an intracellular pool of GLUT1 that also translocates to the plasma membrane in response to insulin. Intracellular stores of GLUT1 and GLUT4 reside in distinct vesicle populations that can be differentiated by centrifugation (Piper *et al.*, 1991a). The glucose transporter compartment could represent a previously unrecognized organelle or it may be a subcompartment of an organelle such as the *trans*-Golgi reticulum. It seems unlikely that a structural element unique to GLUT4 is essential for translocation, given that vesicles containing GLUT1 also translocate to the plasma membrane (Gould *et al.*, 1989). Translocation of glucose transporter vesicles to the cell surface in response to insulin may thus occur independently of the glucose transporters they contain, perhaps as a function of other unique proteins in the organellar membrane.

IV. THE MECHANISM OF GLUT4 TARGETING IS UNKNOWN

Although important progress has been made in the past year concerning the subcellular trafficking of GLUT4, little is known about the mechanism by which this molecule is sorted and how this process is regulated by insulin. The dynamic recycling of GLUT4 (Clark *et al.*, 1991; Slot *et al.*, 1991b) between its intracellular compartment and the plasma membrane has important implications. GLUT4 translocation is often portrayed as the movement of molecules from a static storage compartment to the plasma membrane. This model, which resembles the process of regulated secretion evoked by an exocytic stimulus in neuroendocrine cells, may be an oversimplification considering that endocytosis of cell surface GLUT4 occurs in insulin-stimulated cells (Slot *et al.*, 1991b). The redistribution of GLUT4 in response to insulin may thus be due to changes in its exocytic or endocytic rate constant.

Figure 5 illustrates four possible models of glucose transporter trafficking. The occurrence of GLUT4 endocytosis only after insulin stimulation is shown in Fig. 5a. In this "regulated secretion" model, there is no recycling of GLUT4 in the basal state. This represents the simplest interpretation of the data of Slot *et al.* (1991b). Paradoxically, a very active endocytic process might be impossible to detect by cell surface or immunogold labeling, since very little substrate would be available at the cell surface. Therefore, the data do not exclude GLUT4 recycling in the basal state, as illustrated in Fig. 5b. The key feature of this model is that recycling of GLUT4 occurs continuously between the cell surface and a specialized intracellular membrane compartment. In this "specialized recycling" model, a decrease in the rate of GLUT4 endocytosis or an increase in rate of exocytosis could be an important factor in the insulin response. Another plausible explanation for the inability to detect GLUT4 in the plasma membrane or endocytic pathways under basal conditions is that GLUT4 might be segregated from normal constitutive pathways of exocytosis and endocytosis, but this segregation is abolished by insulin (Fig. 5c). This "constitutive recycling" model does not require a specialized pathway for GLUT4 movement to and from the plasma membrane, but does require movement from a special storage compartment to a constitutive recycling compartment. Finally, in a "salvage" model (Fig. 5d) both transporters initially follow the constitutive pathway of plasma membrane protein flow, but GLUT4 is rapidly endocytosed and targeted to a special insulin-sensitive compartment. In this model, which is reminiscent of the salvaging of resident ER proteins from a pre-Golgi compartment (Pelham, 1989), differential sorting of GLUT1 and GLUT4 occurs in the plasma membrane or at some step after endocytosis. There are, of course, many other equally plausible models. A mechanism incorporating the modulation of GLUT4 exocytosis and endocytosis by insulin would provide the maximum possible degree of control over the subcellular distribution of the protein.

Three other plasma membrane proteins, the gastric H^+/K^+-ATPase (Urushidani and Forte, 1987), the H^+-ATPase of renal intercalated cells (Schwartz and Al-Awqati, 1986) and the water channels of the renal collecting duct (Handler, 1988), are also stored in intracellular compartments and are translocated rapidly in response to a specific stimulus, namely, histamine, antidiuretic hormone, and vasopressin, respectively. A common mechanism may well operate in these diverse cell types. The characterization of other protein constituents of the glucose transporter-containing intracellular compartment is necessary to help answer questions concerning the mechanism of vesicle translocation. The Sec4/Ypt1/rab family of small GTP-binding proteins, related to the p21[ras] superfamily, regulates many steps of intracellular protein transport (Goud and McCaffrey, 1991). Nonhydrolyzable GTP analogs induce accumulation of GLUT4 in the plasma membrane of permeabilized rat adipocytes (Baldini *et al.*, 1991), suggesting that a GTP-binding protein is involved in GLUT4 targeting. Characterization of the GTP-binding protein responsible for this effect and the further exploitation of permeabilized cell systems may yield important insights into the mechanism of GLUT4 redistribution by insulin.

V. CONCLUSIONS

Glucose transporter isoforms are highly homologous but differ markedly in their subcellular targeting. The intracellular sequestration of GLUT4 appears to be independent of cell type, suggesting the involvement of a ubiquitous targeting mechanism. This family of proteins thus provides a

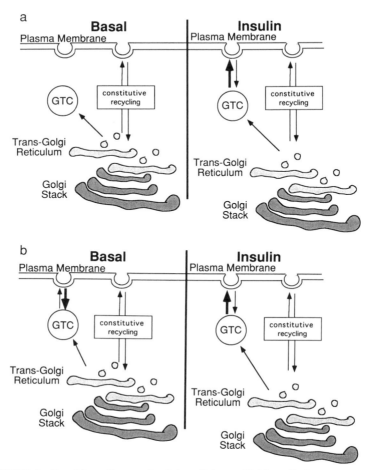

FIGURE 5 Possible pathways for intracellular trafficking of glucose transporters. (a) regulated secretion model; (b) specialized recycling model; (c) constitutive recycling model; and (d) salvage model. GTC, glucose transporter compartment. See text for discussion.

useful model system for studying the sorting of polytopic membrane proteins. An interesting distinction from other sorting problems is that insulin may act to change the kinetics of glucose transporter movement and therefore affect its steady-state distribution. We emphasize the dynamic nature of this process and the possible role of endocytosis in determining the amount of GLUT4 in the plasma membrane. A full understanding of

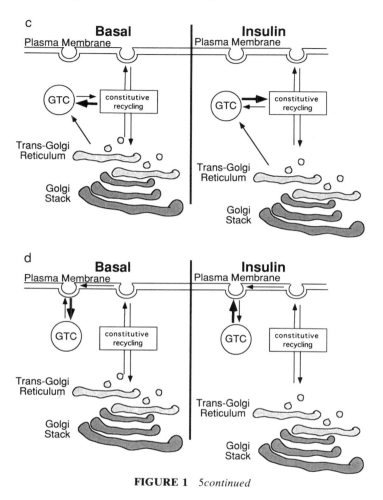

FIGURE 1 *5continued*

the structure–function relationships and intracellular trafficking of glucose transporters may provide insights not only into important medical problems, such as non-insulin-dependent diabetes mellitus (Mueckler and Tordjman, 1991), but also into fundamental concerns in cell biology, such as the nature of signals for intracellular protein targeting and recycling and the mechanisms involved in recognizing and sorting these signals.

References

Adeli, K., and Sinkevitch, C. (1990). Secretion of apoliprotein B in serum-free cultures of human hepatoma cell line, HepG2. *FEBS Lett.* **263**, 345–348.

Baldini, G., Hohman, R., Charron, M. J., and Lodish, H. F. (1991). Insulin and nonhydrolyzable GTP analogs induce translocation of GLUT4 to the plasma membrane in alpha-toxin-permeabilized rat adipose cells. *J. Biol. Chem.* **266**, 4037–4040.

Bell, G., Kayano, T., Buse, J., Burant, C., Takeda, J., Lin, D., Fukumoto, H., and Seino, S. (1990). Molecular biology of mammalian glucose transporters. *Diabetes Care* **13**, 198–206.

Biber, J. W., and Lienhard, G. E. (1986). Isolation of vesicles containing insulin-responsive, intracellular glucose transporters from 3T3-L1 adipocytes. *J. Biol. Chem.* **261**, 16180–16184.

Birnbaum, M. J. (1989). Identification of a novel gene encoding an insulin-responsive glucose transporter protein. *Cell* **57**, 305–315.

Calderhead, D. M., Kitagawa, K., Tanner, L. I., Holman, G. D., and Lienhard, G. E. (1990). Insulin regulation of the two glucose transporters in 3T3-L1 adipocytes. *J. Biol. Chem.* **265**, 13801–13808.

Canfield, W. M., Johnson, K. F., Ye, R. D., Gregory, W., and Kornfeld, S. (1991). Localization of the signal for rapid internalization of the bovine cation-independent mannose-6-phosphate/IGF-II receptor to amino acids 24 to 29 of the cytoplasmic tail. *J. Biol. Chem.* **266**, 5682–5688.

Charron, M. J., de Brosius, F., Alper, S. L., and Lodish, H. F. (1989). A glucose transport protein expressed predominantly in insulin-responsive tissues. *Proc. Natl. Acad. Sci. U.S.A.* **86**, 2535–2539.

Chen, W.-J., Goldstein, J. L., and Brown, M. S. (1990). NPXY, a sequence often found in cytoplasmic tails, is required for coated pit-mediated internalization of the low density lipoprotein receptor. *J. Biol. Chem.* **265**, 3116–3123.

Clancy, B. M., and Czech, M. P. (1990). Hexose transport stimulation and membrane redistribution of glucose transporter isoforms in response to cholera toxin, dibutyryl cyclic AMP, and insulin in 3T3-L1 adipocytes. *J. Biol. Chem.* **265**, 12434–12443.

Clancy, B. M., Harrison, S. A., Buxton, J. M., and Czech, M. P. (1991). Protein synthesis inhibitors activate glucose transport without increasing plasma membrane glucose transporters in 3T3-L1 adipocytes. *J. Biol. Chem.* **266**, 10122–10130.

Clark, A. E., Holman, G. D., and Kozka, I. J. (1991). Determination of the rates of appearance and loss of glucose transporters at the cell surface of rat adipose cells. *Biochem. J.* **278**, 235–241.

Collawn, J. F., Stangel, M., Kuhn, L. A., Esekogwu, V., Jing, S., Trowbridge, I. S., and Tainer, J. A. (1990). Transferrin receptor internalization sequence YXRF implicates a tight turn as the structural recognition motif for endocytosis. *Cell* **63**, 1061–1072.

Conover, C. A., and Lee, P. D. (1990). Insulin regulation of insulin-like growth factor-binding protein production in cultured HepG2 cells. *J. Clin. Endocrinol. Metab.* **70**, 1062–1067.

Cushman, S. W., and Wardzala, L. J. (1980). Potential mechanism of insulin action on glucose transport in the isolated rat adipose cell. Apparent translocation of intracellular transport systems to the plasma membrane. *J. Biol. Chem.* **255**, 4758–4762.

Devaskar, S. U., and Mueckler, M. M. (1992). The mammalian glucose transporters. *Pediatr. Res.* **31**, 1–13.

Ferrannini, E., Bjorkman, O., Reichard, G., Jr., Pilo, A., Olsson, M., Wahren, J., and DeFronzo, A. (1985). The disposal of an oral glucose load in healthy subjects. *Diabetes* **34**, 580–588.

Fukumoto, H., Kayano, T., Buse, J. B., Edwards, Y., Pilch, P. F., Bell, G. I., and Seino, S. (1989). Cloning and characterization of the major insulin-responsive glucose transporter expressed in human skeletal muscle and other insulin-responsive tissues. *J. Biol. Chem.* **264,** 7776–7779.

Gibbs, E. M., Lienhard, G. E., and Gould, G. W. (1988). Insulin-induced translocation of glucose transporters to the plasma membrane precedes full stimulation of hexose transport. *Biochemistry* **26,** 6681–6685.

Goldstein, J. L., Brown, M. S., Anderson, R. G. W., Russel, D. W., and Schneider, W. J. (1985). Receptor-mediated endocytosis: Concepts emerging from the LDL receptor system. *Annu. Rev. Cell Biol.* **1,** 1–39.

Gorman, C. (1985). High efficiency gene transfer into mammalian cells. *In* "DNA Cloning" (D. M. Glover, ed.), Vol. II, pp. 143–190. IRL, Oxford.

Goud, B., and McCaffrey, M. (1991). Small GTP-binding proteins and their role in transport. *Curr. Opin. Cell. Biol.* **3,** 626–633.

Gould, G. W., Derechin, V., James, D. E., Tordjman, K., Ahern, S., Gibbs, E. M., Lienhard, G. E., and Mueckler, M. (1989). Insulin-stimulated translocation of the HepG2/erythrocyte-type glucose transporter expressed in 3T3-L1 adipocytes. *J. Biol. Chem.* **264,** 2180–2184.

Green, H., and Kehinde, O. (1974). Sublines of mouse 3T3 cells that accumulate lipid. *Cell* **1,** 113–117.

Handler, J. S. (1988). Antidiuretic hormone moves membranes. *Am. J. Physiol.* **255,** F375–F382.

Haney, P. M., Slot, J. W., Piper, R. C., James, D. E., and Mueckler, M. (1991). Intracellular targeting of the insulin-regulatable glucose transporter (GLUT4) is isoform specific and independent of cell type. *J. Cell Biol.* **114,** 689–699.

Hirshman, M. F., Goodyear, L. J., Wardzala, L. J., Horton, E. D., and Horton, E. S. (1990). Identification of an intracellular pool of glucose transporters from basal and insulin-stimulated rat skeletal muscle. *J. Biol. Chem.* **265,** 987–991.

Holman, G. D., Kozka, I. J., Clark, A. E., Flower, C. J., Saltis, J., Habberfield, A. D., Simpson, I. A., and Cushman, S. W. (1990). Cell surface labeling of glucose transporter isoform GLUT4 by bis-mannose photolabel. Correlation with stimulation of glucose transport in rat adipose cells by insulin and phorbol ester. *J. Biol. Chem.* **265,** 18172–18179.

Hudson, A. W., Ruiz, M. L., and Birnbaum, M. J. (1992). Isoform-specific subcellular targeting of glucose transporters in mouse fibroblasts. *J. Cell Biol.* **116,** 785–797.

James, D. E., Jenkins, A. B., and Kraegen, E. W. (1985). Heterogeneity of insulin action in individual muscles *in vivo:* Euglycemic clamp studies in rats. *Am. J. Physiol.* **248,** E575–E580.

James, D. E., Lederman, L., and Pilch, P. F. (1987). Purification of insulin-dependent exocytic vesicles containing the glucose transporter. *J. Biol. Chem.* **262,** 11817–11824.

James, D. E., Strube, M., and Mueckler, M. (1989). Molecular cloning and characterization of an insulin regulatable glucose transporter. *Nature (London)* **338,** 83–87.

Johnson, K. F., Chan, W., and Kornfeld, S. (1990). Cation-dependent mannose-6-phosphate receptor contains two internalization signals in its cytoplasmic domain. *Proc. Natl. Acad. Sci. U.S.A.* **87,** 10010–10014.

Kahn, B. B., Simpson, I. A., and Cushman, S. W. (1988). Divergent mechanisms for the insulin resistant and hyperresponsive glucose transport in adipose cells from fasted and refed rats. Alterations in both glucose transporter number and intrinsic activity. *J. Clin. Invest.* **82,** 691–699.

Kahn, B. B., Rosetti, L., Lodish, H. F., and Charron, M. J. (1991). Decreased *in vivo* glucose uptake but normal expression of GLUT1 and GLUT4 in skeletal muscle of diabetic rats. *J. Clin. Invest.* **87**, 2197–2206.

Mueckler, M. (1990). Family of glucose-transporter genes. Implications for glucose homeostasis and diabetes. *Diabetes* **1**, 6–11.

Mueckler, M., and Tordjman, K. (1991). Facilitative glucose transporters: regulation and possible role in NIDDM. *In* "DNA Polymorphisms as Disease Markers" (D. J. Galton and G. Asmann, eds.), pp. 27–41. Plenum, New York.

Mueckler, M., Caruso, C., Baldwin, S. A., Panico, M., Blench, I., Morris, H. R., Allard, W. J., Lienhard, G. E., and Lodish, H. F. (1985). Sequence and structure of a human glucose transporter. *Science* **229**, 941–945.

Munro, S. (1991). Sequence within and adjacent to the transmembrane segment of alpha-2,6-sialyltransferase specify Golgi retention. *EMBO J.* **10**, 3577–3588.

Nilsson, T., Lucocq, J. M., Mackay, D., and Warren, G. (1991). The membrane-spanning domain of beta-1,4-galactosyltransferase specifies trans Golgi localization. *EMBO J.* **10**, 3567–3575.

Pelham, H. R. B. (1989). Control of protein exit from the endoplasmic reticulum. *Annu. Rev. Cell Biol.* **5**, 1–23.

Piper, R. C., Hess, L. J., and James, D. E. (1991a). Differential sorting of two glucose transporters expressed in insulin-sensitive cells. *Am. J. Physiol.* **260**, C570–580.

Piper, R. C., Tai, C., Slot, J., Huang, H., Rice, C., and James, D. E. (1991b). Dissection of the targeting domains of the insulin-regulatable glucose transporter (GLUT4). *J. Cell Biol.* **115**, 303a.

Schwartz, G. J., and Al-Awqati, Q. (1986). Regulation of transepithelial H + transport by exocytosis and endocytosis. *Annu. Rev. Physiol.* **48**, 153–161.

Slot, J. W., Moxley, R., Geuze, H. J., and James, D. E. (1990). No evidence for expression of the insulin-regulatable glucose transporter in endothelial cells. *Nature (London)* **346**, 369–371.

Slot, J. W., Geuze, H. J., Gigengack, S., James, D. E., and Lienhard, G. E. (1991a). Translocation of the glucose transporter GLUT4 in cardiac myocytes of the rat. *Proc. Natl. Acad. Sci. U.S.A.* **88**, 7815–7819.

Slot, J. W., Geuze, H. J., Gigengack, S., Lienhard, G. E., and James, D. E. (1991b). Immuno-localization of the insulin regulatable glucose transporter (GLUT4) in brown adipose tissue of the rat. *J. Cell Biol.* **113**, 123–135.

Smith, R. M., Charron, M. J., Shah, N., Lodish, H. F., and Jarett, L. (1991). Immunoelectron microscopic demonstration of insulin-stimulated translocation of glucose transporters to the plasma membrane of isolated rat adipocytes and masking of the carboxyl-terminal epitope of intracellular GLUT4. *Proc. Natl. Acad. Sci. U.S.A.* **88**, 6893–6897.

Stoorvogel, W., Geuze, H. J., and Strous, G. J. (1987). Sorting of endocytosed transferrin and asialoglycoprotein occurs immediately after internalization in HepG2 cells. *J. Cell Biol.* **104**, 1261–1268.

Suzuki, I., and Kono, T. (1980). Evidence that insulin causes translocation of glucose transport activity to the plasma membrane from an intracellular storage site. *Proc. Natl. Acad. Sci. U.S.A.* **77**, 2542–2545.

Tordjman, K. M., Leingang, K. A., James, D. E., and Mueckler, M. M. (1989). Differential regulation of two distinct glucose transporter species expressed in 3T3-L1 adipocytes: effect of chronic insulin and tolbutamide treatment. *Proc. Natl. Acad. Sci. U.S.A.* **86**, 7761–7765.

Tordjman, K. M., Leingang, K. A., and Mueckler, M. (1990). Differential regulation of the HepG2 and adipocyte/muscle glucose transporters in 3T3L1 adipocytes. Effect of chronic glucose deprivation. *Biochem. J.* **271**, 201–207.

Trowbridge, I. S. (1991). Endocytosis and signals for internalization. *Curr. Opinion Cell Biol.* **3**, 634–641.

Urushidani, T., and Forte, J. G. (1987). Stimulation-associated redistribution of H,K-ATPase activity in isolated gastric glands. *Am. J. Physiol.* **252**, G458–G465.

Yamamoto, Y., Gamble, C. L., Clark, S. P., Joyner, A., Shibuya, T., MacDonald, M. E., Mager, D., Bernstein, A., and Mak, T. W. (1981). Clonal analysis of early and late stages of erythroleukemia induced by molecular clones of integrated spleen focus-forming virus. *Proc. Natl. Acad. Sci. U.S.A.* **78**, 6893–6897.

Zorzano, A., Wilkinson, W., Kotliar, N., Thoidis, G., Wadzinkski, B. E., Ruoho, A. E., and Pilch, P. F. (1989). Insulin-regulated glucose uptake in rat adipocytes is mediated by two transporter isoforms present in at least two vesicle populations. *J. Biol. Chem.* **264**, 12358–12363.

CHAPTER 6

Plasticity in Epithelial Polarity

Qais Al-Awqati, Janet van Adelsberg, and Jiro Takito
Department of Medicine, College of Physicians and Surgeons, Columbia University, New York, New York 10032

I. INTRODUCTION

The plasma membrane of the epithelial cell is differentiated into two domains, apical and basolateral, which differ in composition and function. It is this difference in properties of the two domains that allows the epithelial sheet to perform its major function of transepithelial transport of solutes and water. Transport of these molecules is mediated by membrane proteins that are synthesized on ribosomes that are bound to the endoplasmic reticulum. The nascent proteins then traverse the Golgi cisternae before they are carried to the plasma membrane by small vesicles. During this journey, a decision must be made regarding the ultimate destination of a membrane protein in any cell. The sorting of membrane proteins to various intracellular compartments or to the plasma membrane requires targeting information that is present in the structure of the proteins themselves. Alternatively, the targeting information might be present in the overall organization of the cytoplasm.

Delivery of plasma membrane proteins from their site of synthesis and processing to their destination occurs by transport vesicles that fuse with the plasma membrane, thereby inserting the proteins into that membrane. These exocytotic events can occur in two general ways. In constitutive exocytosis, there is continuous delivery of the proteins to the plasma membrane. In regulated exocytosis, the transport vesicles accumulate under the plasma membrane and only fuse in response to a signal such as an increase in intracellular calcium. Epithelial cells have several specialized pathways for delivery of the proteins to the plasma membranes. Constitutive delivery must be polarized for the cell to achieve its final phenotype. Epithelia also have regulated exocytosis which occurs in a polarized manner. Not all membrane proteins are polarized in epithelial cells, hence another pathway must also exist which "randomly" delivers proteins to apical and basolateral membranes. A final special epithelial process is transcytosis where a membrane protein is carried from the basolateral membrane to the apical surface.

A number of specific protein sequences have been found that target viral or naturally occurring proteins to the apical or basolateral membranes. Removal of such sequences abolishes the targeting and, in some instances, such sequences redirect proteins that are targeted from one membrane to the other. It is likely that these sequences interact with receptors and that these receptors function to target the transport vesicle to its destination. Presumably, the apical or basolateral plasma membrane contains docking proteins which will bind to the targeting receptors of the vesicles. It is customary at this juncture to ask: What targets the targeting receptors to their final destination? Two possibilities come to mind. First, epithelial cells do not arise *de novo,* rather they are the consequence of division of other epithelial cells, *omnis cellula e cellula.* Hence, the daughter cells will contain half of the targeting receptors. In a second possibility, it is clear that epithelial cells can be induced to polarize by interaction with the correct extracellular matrix substrate, hence this asymmetric interaction can induce the polarization of the cytoplasm affecting polarized delivery of the proteins.

An examination of the morphology of an epithelial cell immediately shows that the whole cell is polarized, not just the plasma membrane structures. For instance, the nucleus in epithelial cells is located in the basolateral cytoplasm while the Golgi is located usually in the apical cytoplasm. Detailed analysis of the cytoskeleton shows that the distribution of its many component proteins is polarized. The cytoskeleton is in direct association with cell adhesion molecules and with extracellular matrix proteins. The organization of the cytoskeleton which probably controls the trafficking of vesicles to the apical and basolateral domains is heavily regulated and must control the organization of the cytoplasm. The cell adhesion:cytoskeleton:extracellular matrix network gives the epithelial cell its characteristic shape and might play a dominant role in the targeting of vesicles. It will also likely code for a "maintenance" function where its organized structure might prevent the redistribution and randomization that might occur after a protein has been sorted correctly.

We have been studying the epithelial polarity of the H^+ ATPase and the Cl/HCO_3 exchanger of the intercalated cell of the kidney tubule. This cell exists in two forms with opposite polar distribution of these proteins. Hence, detailed analysis of the process of polarity in the two types of cell might reveal significant insights into the processes that control epithelial polarity in general.

II. TWO TYPES OF INTERCALATED CELL WITH REVERSED POLARITY

Acidification of the urine occurs by the active transport of protons or HCO_3^- by the epithelial cells of the renal tubule. In the cortical collecting tubule, these transepithelial transport processes are mediated by the intercalated cell, one of the two types of cells present in these tubule segments. It is enriched in mitochondria and a large fraction of its volume is occupied by cytoplasmic vesicles. There is overwhelming evidence that this cell is the site of proton and bicarbonate transport. The cell is enriched in carbonic anhydrase, an important component of the H^+/HCO_3 transport system. The intracellular vesicles of the intercalated cells are acidified by proton pumps and these vesicles fuse with the apical or basolateral membranes indicating that the plasma membranes of these cells contain proton pumps. It has also been recently demonstrated that a HCO_3 transport protein, an analogue of the red cell band 3, is also located on the basolateral membrane of one form of these cells. The intercalated cells are present in the connecting tubule and the collecting tubule; they form about a third of the cells.

Based on morphological criteria two forms of these cells have been identified (Table I). The apical membrane of the intercalated cell shows many specializations when examined by the scanning electron microscope (Husted et al., 1981; Madsen and Tisher, 1986); one form has ruffles on its apical surface; while the other has microvilli. These ruffled cells have been termed α-cells while the cells with microvilli were termed β-cells. Recent studies have shown that the α type secretes protons into the urine while the β-cell secretes bicarbonate into the urine (Schwartz et al., 1985). Using freeze-fracture analysis, Stetson and Steinmetz (1985) found that the acid-secreting cells contain intramembrane particles in the apical plasma

TABLE I
Structural, Functional, and Morphological Characteristics of Intercalated Cells

	α Type	β Type
Peanut lectin binding	Absent	Apical
Band 3 (AE1)	Basolateral	Apical
Proton ATPase	Apical	Basolateral
GLUT1	Basolateral	Basolateral
Spectrin	Basolateral	Not known
Band 4.1	Basolateral	Not known
Endocytosis	Apical	Absent
Acid vesicles	Subapical	Subapical and basolateral
CO_2-stimulated exocytosis	Apical	?
Cl conductance	?Basolateral	?Basolateral
Basolateral $\rightarrow$ apical transport	H^+	HCO_3
Apical surface	Microplicae	Microvilli
Dense cytoskeleton	None	Subapical

membrane. Since the β-cells contained these particles on the basolatera
membrane, those authors speculated that these structures might be th
proton-translocating ATPase. Studies by Brown *et al.* (1987) have shown
that these coats are proton pumps since a similar morphology was shown
in purified reconstituted ATPases. It also appears that the intercalated
cell in the cortical collecting tubule has two patterns of staining with
antibodies raised against the purified proton ATPase. In one type, the
ATPase is located largely in the basolateral membrane; this is clearly the
bicarbonate-secreting form of the intercalated cell. In another pattern,
the staining is restricted to the apical and subapical plasma membrane
suggesting that this is the α or acid-secreting phenotype. Finally, some
intercalated cells show a more diffuse staining with the H^+ ATPase antibodies including cytoplasmic staining as well as staining located in the
plasma membrane (Brown *et al.*, 1988). What function this latter type of
cell subserves remains to be identified.

In the rabbit collecting tubule, the bicarbonate-secreting cell has a peanut lectin-binding protein on its apical surface (Le Hir *et al.*, 1982;
Schwartz *et al.*, 1985). This has recently been identified to be a 220-kDa
protein located in the apical membrane (van Adelsberg *et al.*, 1989); its
function is unknown at present. Table I shows the differences between
the two cell types. By using the same transport functions deployed on

different membrane domains these cells exhibit reversed functional polarity, i.e., one secretes protons and the other bicarbonate. Some aspects of their structural polarity is also reversed. Endocytosis is vigorous in the apical membrane of the H^+-secreting cell but is completely absent from the apical membrane of the HCO_3-secreting cell. Increasing the ambient pCO_2 causes apical exocytosis in the H^+-secreting cell but basolateral exocytosis in the HCO_3-secreting cell.

However, not all of the characteristics of the two cells are reversed. For instance, the HCO_3-secreting cell displays a peanut lectin-binding protein that is located on its apical membrane (Le Hir *et al.*, 1982) while the H^+-secreting cell has much less staining for this protein and frequently shows none. The cytoplasm of these two cell types also shows polarity in that the nucleus is located in the basal half of the cell and the tight junctions are near the luminal surface in both types of cells. Although both types of cells show supapical vesicles, the acid-secreting cell demonstrates active fusion events with the apical plasma membrane. In the HCO_3-secreting cell, however, the subapical vesicles are separated from the apical plasma membrane by a dense mesh of cytoskeleton. Hence, when we say that there is reversed epithelial polarity in the two cell types, it should not be taken to mean that the two cells are symmetrically inverted; rather there are obvious and some subtle differences between the two cell types.

III. PLASTICITY IN EPITHELIAL POLARITY

The rabbit cortical collecting tubule secretes HCO_3 into the lumen. When rabbits are fed an acid diet for a few days the direction of HCO_3 movement is reversed to HCO_3 absorption, i.e., H^+ secretion. Since the transport rates that are measured are net values in a tubule that contains both types of cells there were two possible interpretation of this finding; one interpretation is that the capacity of the H^+-secreting cell to secrete protons is increased much greater than the amount of HCO_3 added to the lumen by the other cell type. The other possibility is that the HCO_3-secreting cell reversed its epithelial polarity to become a H^+-secreting cell. To distinguish between these two possibilities we counted the two cell types in control and in acid-fed animals. The animals were fed NH_4Cl by stomach tube and it was found that although their plasma HCO_3 concentration decreased initially to low levels, this value recovered back to the original level within less than a day. Cortical collecting tubules were then dissected from control and acid-fed animals and the number of H^+-secreting and HCO_3-secreting cells was measured. The total number of intercalated cells was the same in control and acid-fed rabbits, but the number of H^+-secreting cells increased by a factor of 10, while the number of HCO_3-secreting cells decreased by the same number. These results are most compatible with conversion of epithelial polarity. However, they do not rule out the possibility that the H^+-secreting cells proliferated while the HCO_3-secreting cells died and that the two processes occurred with sufficiently similar rates to account for a constant total number of intercalated cells. Since a 10-fold increase in the number of H^+-secreting cells within 1–4 days should have been accompanied by an increased appearance of mitotic figures, we assayed the frequency of mitotic figures in the collecting tubules and found no difference between control tubules and those removed from acid-treated animals. In fact there were no mitotic figures in any cell type in these tubules even though we would have predicted 5–10/tubule. It is on the basis of these experiments that we concluded that there was plasticity in epithelial polarity.

The most direct method to demonstrate that there is actual conversion of polarity is to start with a pure population of cells that displays the polar distribution and the plasticity in epithelial polarity. The HCO_3-secreting intercalated cell is characterized by the presence of a peanut lectin-binding site on its apical surface. In rabbit kidney cortex, unlike the kidneys of other animals, the HCO_3-secreting intercalated cell is the only cell that binds peanut lectin. We developed a cell purification scheme that started with collagenase digestion followed by plating cells on large petri dishes that have been coated with peanut lectin (van Adelsberg *et al.*, 1989). Bound cells were eluted with 300 mM galactose and purified further on Ficoll density gradient. The purified cells were plated on collagen-coated filters and grown in serum-free media. Morphological studies at the light and electron microscopic levels showed that the cells had obvious tight junctions, basolateral infolding, abundant mitochondria, and apical peanut lectin staining. The monolayers had a high resistance of about 600 ohms · cm^2. There was transepithelial HCO_3 secretion in the absence of basolateral HCO_3 which was reduced to near zero when the apical Cl was removed. In the presence of basolateral HCO_3 the Cl-dependent flux increased.

To obtain enough material for biochemical and molecular studies, we immortalized the purified β-cells using transfection with a temperature-sensitive mutant of the SV40 large T antigen. This was cotransfected with a neomycin resistance gene into a purified HCO_3-secreting intercalated cell using electroporation. Four resistant clones grew at the permissive temperature (33°C) and when they were shifted to the nonpermissive temperature of 40°C, the expression of the large T antigen started to disappear as assayed by immunocytochemistry. At the nonpermissive temperature, the cells expressed apical peanut lectin binding and had tight junctions by electron microscopy. Confluent monolayers of these cells had no endocytosis from the apical medium. The resistance of these mono- layers was too low to allow measurement of transepithelial HCO_3^- secre- tion. We demonstrated that these cells contained an apical Cl/HCO_3 ex- changer by measurement of the effect of changes in the apical or basolateral concentrations of Cl on intracellular pH (Edwards *et al.*, 1992). Hence, these cells retain all the important characteristics of the β inter- calated cells.

IV. THE APICAL AND BASOLATERAL Cl/HCO₃ EXCHANGERS ARE BOTH AE1

The acid-secreting intercalated cell transports HCO_3 across the basolat- eral membrane via a band 3-like protein. Three band 3-related genes have been identified which are homologous to each other in the membrane- spanning C-terminal domains but are quite divergent at the N-terminal region (Kopito, 1990). Numerous studies largely based on specific antibod- ies raised against the N-terminal region of AE1 have suggested that the basolateral protein is similar to the erythroid form of this Cl/HCO_3 ex- changer. To identify the molecular nature of these band 3 genes we pre- pared mRNA from the purified and immortalized intercalated cells. We screened mRNA from the cells with a probe for the membrane-spanning domain and found that there were two bands of 4.6 and 5.0 kb. When the blot was screened using probes specific for the erythroid (AE1) and the nonerythroid (AE2) forms, we found that they hybridized to the individual bands with the erythroid, one being the 5 kb and the nonerythroid the 4.6 kb. However, the erythroid form was at least 10 times more abundant than the nonerythroid form in both types of cells. These results suggest that the apical Cl/HCO_3 exchanger is likely to be a product of the erythroid form of band 3.

We prepared three classes of antibodies against various regions of the erythroid band 3, one against the C-terminal peptide, one against the membrane-spanning domain, and another against the N-terminal region. The antibody against the membrane-spanning region should cross-react with the nonerythroid form while the others are more specific for the erythroid form. Using these antibodies we found that they identified an erythroid form of the band 3 protein in membranes of immortalized HCO_3^-- secreting intercalated cells. To demonstrate the location of the band 3 homologue, we used the cell-ripping method devised by Sambuy *et al.* In the method the apical membrane of confluent monolayers is coated with silica and fused to glass slides (Sambuy and Rodriguez–Boulan, 1988). This separates the apical membranes from the rest of the cell. Using this method, we found that the peanut lectin binding was restricted to the apical fraction, as expected, and that GLUT1 was restricted to the basolateral fraction. These control studies suggest that this method can separate the apical from basolateral fraction. Using this method, we used an antibody raised against the N-terminus of rabbit erythrocyte band 3. This affinity-purified antibody recognized a 100-kDa protein in the apical fraction but none in the basolateral fraction. These results demonstrate that the apical Cl/HCO_3 exchanger is AE1 (van Adelsberg *et al.*, 1993).

Is band 3, the product of the AE1 gene, the polypeptide identified by our antibody? Two lines of evidence suggest that this protein is in fact AE1, i.e., band 3. First, our antibody is specific for the cytoplasmic domain of band 3. Although the cDNA's for rabbit AE1 and AE2 have not been sequenced, data from the mouse suggest that the cytoplasmic domains of these proteins will be about 33% homologous at the amino acid level (Kopito, 1990). This contrasts to the >85% homology of the membrane-spanning domains of AE1, AE2, and AE3 (Kopito, 1990). Therefore, our antibody is likely to be specific for band 3. Second, the polypeptide that we have identified is 100 kDa, an appropriate size for AE1 proteins. The AE2 gene product has a predicted molecular mass of 137 kDa; when it is expressed in COS cells it migrates at a molecular mass of 140 and 165 kDa in polyacrylamide gels (Lindsey *et al.*, 1990). AE2 is the only known anion exchanger expressed in choroid plexus; Western blots of this tissue probed with an antibody raised against a C-terminal peptide common to both AE1 and AE2 recognized a polypeptide of 165 kDa (Lindsey *et al.*, 1990). These data suggest that the AE2 gene product is >140 kDa in molecular weight. Finally, we have cloned a large fragment of the intercalated cell AE2 and expressed it as a fusion protein; the AE1 specific antibody did not react with it on immunoblots. Therefore, the band 3-related polypeptide that we have identified is the product of the AE1 gene.

These observations immediately raise the question of how the polarity of band 3 in α and β intercalated cells is generated and maintained. The α intercalated cells have basolateral band 3 that is indistinguishable, at least at the biochemical level, from the band 3 that is sorted to the apical plasma membrane in β intercalated cells. The least likely explanation is that the basolateral form of the protein is not band 3 but is the product of the AE2 gene expressed in kidney. This explanation seems extremely unlikely as the basolateral form of the protein is recognized by antipeptide antibodies raised against peptides that are unique to AE1 (Alper et al., 1989). A second hypothesis is that polarities of both band 3 and the cytoskeleton of β intercalated cells are reversed. Band 3 has well-characterized interactions with the cytoskeleton in erythrocytes and colocalizes with ankyrin and spectrin in the basolateral membrane of α intercalated cells. Na,K-ATPase binds to anykrin; this binding causes basolateral localization of the enzyme in Madin–Darby canine kidney (MDCK) by increasing its half-life (Nelson and Hammerton, 1989; Nelson et al., 1990). In retinal pigment epithelial cells, reversal of the polarity of the cytoskeleton is associated with reversal of the polarity of Na,K-ATPase expression (Gunderson et al., 1991). Erythrocyte and brain isoforms of ankyrin are both polarized and differentially expressed in kidney epithelial cells (Davis et al., 1989). Examination of these ankyrin isoforms in the two types of intercalated cells might reveal differences in their expression or localization. A third hypothesis is that the sorting signal resides in a subtle post-translational modification of band 3. Band 3 in erythrocytes is phosphorylated; this phosphorylation inhibits band 3 cytoplasmic domain binding to glycolytic enzymes and hemoglobin (Low et al., 1987). Band 3 has also been reported to be palmitoylated (Okubo et al., 1991).

These observations also raise significant questions about the function of band 3 in apical and basolateral plasma membrane domains. The apical Cl/HCO_3 exchanger has a higher K_M for Cl^-, lower 4,4[1],–diisothiocyano–2,2[1]–stilbene disulfonate (DIDS) sensitivity and it is not recognized by antibodies in immunocytochemical experiments.

We have preliminary evidence which suggests that the differences in the kinetics of activation of Cl/HCO_3 exchange by Cl and the DIDS inhibition of apical and basolateral kidney band 3 might be due to the lipid composition of the membrane rather than to an intrinsic characteristic of the band 3 protein. We found that when erythrocyte band 3 was reconstituted into different lipids, the K_m and V_{max} for Cl^- in rabbit erythrocyte as well as the DIDS sensitivity was altered. Interestingly, lipids like sphingomyelin and gangliosides, which are characteristically found in the apical plasma membrane (Simons and van Meer, 1988), decrease V_{max} and increase the K_m for Cl^- of erythrocyte band 3. This observation suggests that the basis for the difference in function may lie in the different lipid environments of the apical and basolateral membranes. The reason why band 3 does not expose any of its cytoplasmic epitopes while in the apical membrane remains obscure. The most likely explanation is that the subapical cytoskeletal network is sufficiently dense to prevent the access of the antibody. When the components of this network become known, one will be able to test this hypothesis directly.

V. PLASTICITY OF EPITHELIAL POLARITY *IN VITRO*

We first demonstrated that both the primary and immortalized inter-calated cell have all the characteristics of the HCO_3-secreting phenotype. These include an apical peanut lectin-binding protein, Cl/HCO_3 exchange at the apical border, and no endocytosis from the apical membrane. To demonstrate reversal of functional polarity in these cells we should recall that the acid-secreting intercalated cell has basolateral Cl/HCO_3 exchange, no apical peanut lectin binding, and active apical endocytosis.

We found that growth conditions could induce this plasticity *in vitro*. We found that when HCO_3-secreting intercalated cells were seeded at subconfluent densities ($10^5/cm^2$) and allowed to become confluent they grew to become HCO_3-secreting intercalated cells which had all the charac-teristics of the phenotype mentioned above. However when they were seeded at superconfluent densities ($>10^6/cm^2$) they grew to become acid-secreting intercalated cells. These cells had apical endocytosis, expressed basolateral band 3, and lost apical peanut lectin binding. Acid-secreting intercalated cells (i.e., cells grown at high density) internalized fluorescent dextran actively. However, the HCO_3-secreting intercalated cells, i.e., cells grown at low density, had no uptake of endocytic markers. The same band 3 message was expressed in both types of cells. Preliminary results suggest that the same band 3 protein is produced in both types.

What are the potential mechanisms by which density of seeding could lead to a change in phenotype? One possibility is that increased cell to cell contact could induce the secretion of some extracellular matrix component that determines the form of polarity. To test this hypothesis, we seeded cells at high density and allowed them to grow till confluence, we then carefully solubilized the cells with detergent and plated cells on the filters at low density and allowed them to grow to confluence. These cells grown at low density now had the acid-secreting phenotype. Control experiments in which cells were first grown at low density and solubilized and new cells seeded also at low density retained the HCO_3-secreting phenotype. These results demonstrate that a material is deposited on the filters that has the capability of reversing the polarity of the β-cell into an α-cell. Cells were incubated with [^{35}S]methionine and plated at high or low density and their extracellular matrix (ECM) was analyzed. The ECM extracted from cells plated at superconfluent densities contained a new 230-kDA protein. This protein was produced during the first day after plating and had a long half-life. Hence, it has the potential for being a "molecular switch." When cells were plated on a semipurified 230-kDa protein they converted their polarity as judged by the development of apical endocytosis.

VI. POTENTIAL MECHANISMS BY WHICH ECM PROTEINS DETERMINE POLARITY OF INTERCALATED CELLS

Extracellular matrix proteins are well known to control epithelial polarity. Such proteins must bind to receptors on the basolateral membrane. Epithelial cells grown in suspension are not polarized. But plating such cells on extracellular matrix induces polarization. Further, when ECM proteins are added to the apical side of polarized epithelia they induce reorientation of the epithelial monolayers such that the apical side becomes basolateral. The dominant role that ECM proteins play in redirecting the polar distribution of apical and basolateral membrane proteins has been documented in a number of *in vitro* cells such as MDCK cysts, embryonic trophectoderm, and thyroid epithelial follicles. In each instance, the reversal of polarity induced in these cells, however, should be distinguished from the one we have described for the intercalated cells. When these systems reverse their polarity, the whole cell reverses its polarity. The tight junctions open and are transferred to the other side; all previous apical proteins move to the other side and there is a complete reversal of the cell orientation. Let us term the role that ECM proteins play here as a *generic* role.

In the intercalated cell, only some proteins are reversed in polarity. The glucose transporter, GLUT1, is basolateral in both cell types; the tight junction is apical in both types; the nucleus is in the basolateral half of the cell in both types. Only some membrane and cytoskeletal proteins are reoriented. We term the role of the 230-kDA ECM protein as *specific* rather than generic.

That band 3 can be retargeted from one domain to another has a number of implications for epithelial polarity. First, band 3 might be an example of a class of proteins that do not in themselves bear a dominant targeting signal. Rather, the signals that it has follow the targeting of its cytoskeletal-binding proteins. The most likely scenario for band 3 is that its cytoplasmic domain binds to cytoskeletal proteins. During the polarity conversion, it is likely that the orientation of the cytoskeleton will change and band 3 will be trapped in the basolateral membrane. Some studies have suggested that Na,K-ATPase also binds to ankyrin and might also lack specific targeting signals (Hammerton *et al.*, 1991). For instance, retinal pigment epithelium has an apical distribution of Na,K-ATPase and ankyrin *in situ*. However, when the cells are cultured *in vitro*, the protein seems to lose its specific polarity (Nabi *et al.*, 1993). Hence, we can conceive of two classes of membrane proteins, those which have a dominant targeting sequence that guides them to their destination and those that do not have specific targeting signals but rather are retained and trapped in one membrane domain. Trapping implies that the proteins bind to cytoskeletal elements; hence it is the organization of the cytoskeleton that is the primary event here. It seems that the organization of the cytoplasm is plastic and can control the redistribution of specific membrane proteins without loss of polarity. Interaction of epithelial cells with ECM has been thought to induce polarity when no polarity existed. That a single new protein can reverse polarity of some without altering the fact of polarity or the polar distribution of other proteins implies that the details of specific protein targeting can be regulated by synthesis or degradation of specific ECM proteins or their receptors.

ECM proteins bind to basolateral receptors which belong to the integrin family. Such receptors have a cytoplasmic tail that interacts with the cytoskeleton via a number of proteins. The affinity of the receptor to the ECM proteins and to the cytoskeleton is known to be regulated by a number of second messengers acting through kinases and phosphatases. Hence, the potential for a specific new ECM protein induced by cell density to regulate the orientation of the cytoskeleton exists. The organization of the cytoskeleton then depends on the following set of proteins: ECM proteins, ECM receptors, cytoplasmic-binding proteins to the receptor, and finally the cytoskeletal proteins themselves. Each of these classes of proteins is regulated in terms of number and affinity by proteases kinases and phosphatases.

Retargeting of band 3 from apical to basolateral domains of the intercalated cell requires reorganization of the subapical cytoskeletal network. This must be coordinated with the basolateral reorganization of the cytoplasm. Since band 3 binds to ankyrin and spectrin, an important question to ask is whether the apical cytoskeleton of the β-cell contains ankyrin, or whether band 3 is retained there by interaction with other cytoskeletal proteins. Further, this cytoskeleton must be disaggregated before H^+ ATPase vesicles can fuse with the apical membrane. What are the potential mechanisms that would allow disaggregation of such a dense network of cytoskeleton? Activation of specific proteases is a likely event which might then allow the cell to start to reorganize its newly synthesized cytoskeleton into the α type of cytoskeleton. In the β type of intercalated cell, band 3 is located in the apical membrane, and its cytoplasmic tail is buried in the dense cytoskeleton that is in the subapical region. The proteins that compose this dense mat are unknown at present. Why does the band 3 go to the apical membrane, or, based on the above discussion, why does absence of the 230-kDa ECM protein result in organization of the apical cytoskeleton in such a way as to trap band 3? Is this the default pathway? Or is there a protein whose role is equivalent to the 230-kDa basolateral protein ECM? One interesting possibility is that reorganization of the cell is accompanied by degradation of the peanut lectin-binding protein. Perhaps that protein is the agent for organization of the apical cytoskeleton.

It is clear from this brief discussion that retargeting of band 3 in these cells is a richly complex subject with ramifications for several areas in epithelial cell biology.

References

Alper, S. L., Natale, J., Gluck, S., Lodis, H. F., and Brown, D. (1989). Subtypes of intercalated cells in rat kidney collecting duct defined by antibodies against erythroid band 3 and renal vacuolar H^+-ATPase. *Proc. Natl. Acad. Sci. U.S.A.* **86**, 429–433.

Brown, D., Gluck, S., and Hartwig, J. (1987). Structure of the novel membrane-coating matrial in proton-secreting epithelial cells and identification as an H^+ATPase. *J. Cell Biol.* **105**, 1637–1648.

Brown, D., Hirsch, S., and Gluck, S. (1988). An H^+ATPase is present in opposite plasma membrane domains in subpopulations of kidney epithelial cells. *Nature (London)* **331**, 622–624.

Caplan, M. J., Anderson, H. C., Palade, G. E., and Jamieson, J. D. (1986). Intracellular sorting and polarized cell surface delivery of (Na^+, K^+) ATPase, an endogenous component of MDCK cell basolateral plasma membranes. *Cell* **46**, 623–631.

Davis, J. Q., Davis, L., and Bennett, V. (1989). Diversity in membrane binding sites of ankyrins. *J. Biol. Chem.* **264**, 6417–6426.

Edwards, J. C., van Adelsberg, J., Rater, M., Herzlinger, D., Lebowitz, J., and Al-Awqati, Q. (1992). Conditional immortalization of bicarbonate-secreting intercalated cells from rabbit. *Am. J. Physiol.* **263**, C521–C529.

Gunderson, D., Orlowski, J., and Rodriguez-Boulan, E. (1991). Apical polarity of Na^+, K-ATPase in retinal pigment epithelium is linked to a reversal of the ankyrin–fodrin submembrane cytoskeleton. *J. Cell Biol.* **112**, 863–872.

Hammerton, R. W., Krzeminski, K. A., Mays, R. W., Ryan, R. A., Wollner, D. A., and Nelson, W. J. (1991). Michanism for regulating cell surface distribution of Na^+, K^+ATPase in polarized epithelial cells. *Science* **254**, 847–850.

Husted, R., Mueller, A., Kessel, R. G., and Steinmetz, P. R. (1981). Surface characteristics of carbonic-anhydrase-rich cells in turtle urinary bladder. *Kidney Int.* **19**, 491–502.

Kopito, R. R. (1990). Molecular biology of the anion exchanger gene family. *Int. Rev. Cytol.* **123**, 177–199.

Le Hir, M., Kaissling, B., Koeppen, B. M., and Wade, J. B. (1982). Binding of peanut lectin to specific epithelial cell types in kidney. *Am. J. Physiol.* **242**, C117–120.

Lindsey, A. E., Schneider, K., Simmons, D. M., Baron, R., Lee, B. S., and Kopito, R. R. (1990). Functional expression and subcellular localization of an anion exchanger cloned from choroid plexus. *Proc. Natl. Acad. Sci. U.S.A.* **87**, 5278–5282.

Low, P. S., Allen, D. P., Zioncheck, T. F., Chari, P., Willardson, B. M., Geahlen, R. L., and Harrison, M. L. (1987). Tyrosine phosphorylation of band 3 inhibits peripheral protein binding. *J. Biol. Chem.* **262**, 4592–4596.

Madsen, J. M., and Tisher, C. C. (1986). Structural–functional relationships along the distal nephron. *Am. J. Physiol.* **250**, F1–F15.

Nabi, I., Mathews, A. P., Cohen-Gould, L., Gunderson, D., and Rodriguez-Boulan, E. (1993). Immortalization of polarized retinal pigment epithelial cells. *J. Cell. Sci.* **104**, 37–49.

Nelson, W. J., and Hammerton, R. W. (1989). A membrane-cytoskeletal complex containing Na$^+$, K$^+$-ATPase, ankyrin, and fodrin in madin–darby canine kidney (MDCK) cells: implications for the biogenesis of epithelial cell polarity. *J. Cell Biol.* **108,** 893–902.

Nelson, W. J., Shore, E. M., Wang, A. Z., and Hammerton, R. W. (1990). Identification of a membrane–cytoskeletal complex containing the cell adhesion molecule uvomorulin (E-cadherin), ankyrin, and fodrin in madin-darby canine kidney, epithelial cells. *J. Cell Biol.* **110,** 349–357.

Okubo, K., Hamasaki, N., Hara, K., and Kageura, M. (1991). Palmiloylation of cysteine 69 from the COOH-terminal of band 3 protein in the human erythrocyte membrane acylation occurs in the middle of the consensus sequence of F–I–IICLAVL found in band 3 protein and G2 protein of rift valley fever virus. *J. Biol. Chem.* **266,** 16420–16424.

Sambuy, Y., and Rodriguez-Boulan, E. (1988). Isolation and characterization of the apical surface of polarized madin–darby canine kidney epithelial cells. *Proc. Natl. Acad. Sci. U.S.A.* **85,** 1529–1533.

Schwartz, G. J., Barasch, J., and Al-Awqati, Q. (1985). Plasticity of functional epithelial polarity. *Nature (London)* **318,** 368–371.

Simons, K., and van Meer, G. (1988). Lipid sorting in epithelial cells. *Biochemistry* **27,** 6197–6202.

Stetson, D. L., and Steinmetz, P. R. (1985). A and b types of carbonic anhydrase-rich cells in turtle bladder. *Am. J. Physiol* **249,** F553–F565.

van Adelsberg, J., Edwards, J. C., Herzlinger, D., Cannon, C., Rater, M., and Al-Awqati, Q. (1989). Isolation and culture of HCO$_3$-secreting intercalated cells. *Am. J. Physiol.* **256,** C1004–C1011.

van Adelsberg, J. S., Edwards, J. C., and Al-Awqati, Q. (1993). The apical Cl/HCO$_3$ exchanger of β intercalated cells. *J. Biol. Chem.,* **268,** 11283–11289.

CHAPTER 7

Regulation of Cell Adhesion and Development of Epithelial Cell Surface Polarity

W. James Nelson
Department of Molecular and Cellular Physiology, Stanford University School of
Medicine, Stanford, California 94305

I. PERSPECTIVE

Cell–cell interactions in development are controlled by the expression of specific cell adhesion proteins. Following cell adhesion, cells undergo a complex remodeling process to form multicellular organs and tissues with distinct functions. Our understanding of the interrelationship between cell adhesion and cell morphogenesis, and the mechanisms that lead to cellular remodeling and differentiation, is relatively poor. To approach these problems, we have been studying a model system of renal transporting epithelial cells, Madin–Darby canine kidney (MDCK) cells. In the absence of cell–cell contact MDCK cells have a relatively poor degree of cellular polarity. However, activation of the cell adhesion protein E-cadherin results in formation of specific cell–cell contacts and subsequent remodeling of cells into a monolayer of polarized transporting epithelial cells. Here, the roles of cell–cell adhesion through E-cadherin, assembly of the membrane cytoskeleton, and membrane protein sorting pathways are discussed in the context of mechanisms that may be involved in this cellular remodeling process.

II. REGULATION OF CELL ADHESION

Multicellular organisms are comprised of heterogeneous cell types that are organized during development into distinct patterns to form tissues and organs. One of the most important primary processes involved in regulating the establishment and maintenance of these cell patterns is cell–cell adhesion (Trinkaus, 1984). Pioneering studies by Holtfreter and colleagues, Moscona, and Steinberg (reviewed in Trinkaus, 1984) established a central principle that the interaction between cells within a heterogeneous cell population was based on the specificity and extent of adhesion between those cells; i.e., cells of the same type tended to aggregate together and, therefore, sorted-out from other cells. The molecular basis for cell adhesion has more recently been shown to be due to the cell surface expression of a family of glycoproteins that bind with high specificity to each other on adjacent cells (reviewed in Edelman and Crossin, 1991; Takeichi, 1990,1991). Significantly, these proteins are expressed in distinct patterns during tissue and organ morphogenesis suggesting that they play an important and direct role in the temporal and spatial regulation of cell interactions and cell sorting during tissue formation (reviews in Edelman and Crossin, 1991; Takeichi, 1988). Furthermore, loss of expression of these proteins correlates with loss of intercellular adhesion, which is an early event in the induction of metastatic disease (Behrens *et al.,* 1989; Shimoyama *et al.,* 1989; Vleminckx *et al.,* 1991).

Two functionally distinct mechanisms of cell adhesion have been described, Ca^{2+}-dependent and Ca^{2+}-independent adhesion, both of which are mediated by a growing number of cell surface glycoproteins (reviewed in Edelman and Crossin, 1991; Takeichi, 1990). These proteins have been classified into three major families of cell adhesion molecules (CAMs): the immunoglobulin (Ig) superfamily, integrins, and cadherins. N-CAM is one of the most thoroughly characterized members of the Ig superfamily of CAMs (Cunningham *et al.,* 1987; Edelman and Crossin, 1991). N-CAM interactions on adjacent cells are homotypic and are Ca^{2+}-independent. Early studies on purified N-CAM that was reconstituted into liposomes provided direct biochemical evidence of modulation of CAM avidity by CAM concentration; a 2-fold increase in the amount of N-CAM produced a 30-fold increase in the binding rate (Hoffman and Edelman, 1983). The integrin family of CAMs is mostly involved in cell interactions with the substratum. However, LFA-1, a member the integrin family expressed on lymphocytes, interacts with I-CAM, a ligand on the surface of endothelial cells (Springer, 1990).

III. THE CADHERIN FAMILY OF CELL ADHESION PROTEINS

The cadherin family of CAMs mediates Ca^{2+}-dependent cell–cell adhesion in a wide variety of cell types (reviewed in Takeichi, 1990;1991; Takeichi, 1988). Three basic classes of cadherins have been characterized, E-, P-, and N-cadherin, although the number of distinct proteins in the cadherin family is increasing almost daily. Initial evidence that cadherins play a role in modulating cell interactions was obtained using antibodies raised against the extracellular domain of a given cadherin, which was released from cells by trypsin in the presence of Ca^{2+} (Behrens et al., 1985; Hatta et al., 1985; Vestweber and Kemler, 1984; Vestweber et al., 1985). These antibodies inhibit cell–cell adhesion and cause clusters of cells to disaggregate in tissue culture (Gumbiner et al., 1988). Significantly, addition of specific cadherin antibodies to preimplantation mouse embryos blocks compaction (Hyafil et al., 1981; Vestweber and Kemler, 1984), and their addition to developing tissues in the embryo inhibits further development (Hirai et al., 1989; Vestweber and Kemler, 1984); similar results have been obtained with antibodies to other CAMs (e.g., N-CAM; Gallin et al., 1986).

Molecular analysis of cadherin structure–function relationships has provided important insights into the role of different protein domains in homotypic binding between cadherins, and interactions between cadherins and cytoplasmic proteins (reviewed in Edelman and Crossin, 1991; Takeichi, 1988). Cadherins span the membrane once with their N-terminal located in the extracellular space and C-terminal in the cytoplasm (Fig. 1). Different cadherins exhibit an overall sequence homology of 43–58% (Takeichi, 1988). The extracellular domain of different cadherins exhibits the least homology, while the cytoplasmic domain has the highest homology. The extracellular domain comprises three repeat domains, each of which contains two putative Ca^{2+}-binding sites (Ringwald et al., 1987) (Fig. 1). Mutagenesis of an amino acid in one of the Ca^{2+}-binding sites in the N-terminal repeat domain abolishes cell adhesion (Ozawa et al., 1990a).

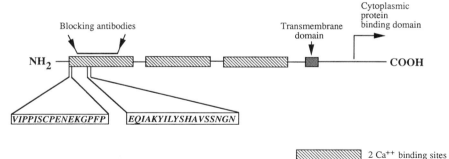

FIGURE 1　Schematic representation of the functional domains of the cell adhesion protein E-cadherin. (Data compiled from Ringwald *et al.*, 1987; Blaschuk *et al.*, 1990; Nagafuchi *et al.*, 1987; Nagafuchi *et al.*, 1987; Ozawa *et al.*, 1989,1990b).

Recent evidence suggests that the N-terminal repeat contains a cell adhesion recognition motif. The N-terminal 113 amino acid domain contains a tripeptide, *HAV*, that is highly conserved between cadherin classes (Fig. 1). Antibodies raised against this domain inhibit cadherin function and block cell adhesion (Behrens *et al.*, 1985; Damsky *et al.*, 1983; Hatta *et al.*, 1985; Vestweber and Kemler, 1984,1985; Vestweber *et al.*, 1985). Significantly, synthetic peptides containing the *HAV* tripeptide inhibit compaction of mouse preimplantation embryos (Blaschuk *et al.*, 1990), which normally occurs through E-cadherin-induced cell adhesion (Vestweber and Kemler, 1984). These results indicate that the N-terminal repeat domain of cadherins may be important for homotypic binding function. However, little is known about the affinity of these interactions, or the role of the cytoplasmic domain in modulating the avidity of homotypic E-cadherin binding.

Definitive evidence that cadherins are involved in homotypic binding in cell–cell adhesion was obtained when cadherin cDNAs were expressed in fibroblasts (Nagafuchi *et al.*, 1987; Mege *et al.*, 1988; McNeill *et al.*, 1990). Fibroblasts do not exhibit significant Ca^{2+}-dependent cell adhesion properties. However, expression of cadherins in these cells results in Ca^{2+}-dependent cell–cell recognition and the formation of compact cell aggregates (colonies) similar to those formed by epithelial cells (Fig. 2). Significantly, mixtures of cells expressing different classes of cadherins sort out from one another such that cells expressing the same cadherin class, or similar cell surface levels of cadherin, form separate aggregates (Hatta *et al.*, 1988; Mege *et al.*, 1988).

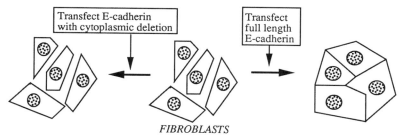

FIGURE 2   Fibroblast expression assay to determine functional domains of E-cadherin in cell–cell adhesion and attachment to cytoplasmic (cytoskeletal) proteins. (Data compiled from Nagafuchi *et al.*, 1987; Mege *et al.*, 1988; McNeill *et al.*, 1990.)

Through studies of cadherin function in this fibroblast expression system, evidence has also been obtained that the cytoplasmic domain of cadherins plays an important role in cell adhesion. Deletions of the C-terminal half of the cytoplasmic domain result in the expression of a truncated protein at the cell surface with an intact extracellular domain (Nagafuchi and Takeichi, 1988; Ozawa *et al.*, 1989,1990b). However, these cells do not form stable cell–cell contacts, or localize cadherin to cell–cell contacts, even when the cells are grown at high density (Fig. 2). The loss of cadherin function as a result of these cytoplasmic deletions appears to be due, at least in part, to loss of interactions with cytoplasmic proteins (Nagafuchi and Takeichi, 1988; Ozawa *et al.*, 1989,1990b; McNeill *et al.*, 1990).

IV. INTERACTIONS BETWEEN CADHERINS AND CYTOPLASMIC (CYTOSKELETAL) PROTEINS

Indirect evidence that cadherins associate with cytoplasmic proteins has come from the observation that cadherins colocalize with actin-associated junctions to sites of cell contact (Hirano *et al.*, 1987). In addition, a limited number of cytoplasmic proteins have been found to coimmunoprecipitate with cadherins solubilized from whole cells (Nagafuchi and Takeichi, 1988; Ozawa *et al.*, 1989,1990b). The most prominent of these cytoplasmic proteins have apparent molecular weights of 102,000, 92,000, and 84,000, and have been termed α-, β-, and γ-catenin, respectively (Ozawa *et al.*, 1989,1990b).

MDCK cells contain several cytoplasmic proteins that appear to bind to E-cadherin. These proteins were identified by coimmunoprecipitation and by isolation of protein complexes in sucrose gradients and nondenaturing polyacrylamide gels. Immunoprecipitation of E-cadherin from MDCK cell extracts results in the precipitation of E-cadherin ($M_r \sim 120,000$), three lower molecular weight proteins ($M_r \sim 102,000, 92,000$, and $84,000$; Shore and Nelson, 1991) that have electrophoretic mobilities similar to those of the α-, β-, and γ-catenins that have been described in other cell types (Ozawa *et al.*, 1990b), and ankyrin and fodrin (Nelson *et al.*, 1990b; Shore and Nelson, 1991). These proteins are not exposed to the cell surface since they are not labeled by lactoperoxidase-catalyzed cell surface iodination (Shore and Nelson, 1991). We have undertaken preliminary analysis of the interactions of these cytoplasmic proteins with E-cadherin by subjecting the immunoprecipitates to buffers of different ionic strengths and pH; at 500 mM, salt, ankyrin, and fodrin are lost from the complex; at pH 2.5, there is a quantitative loss of the M_r 92,000 and 84,000 proteins and ankyrin and fodrin, but retention of the M_r 102,000 protein in a complex with E-cadherin (Piepenhagen and Nelson, 1993; see also McCrea *et al.*, 1991b; Ozawa *et al.*, 1990b).

The nucleotide sequence of α-catenin (M_r 102,000) has recently been published (Nagafuchi *et al.*, 1991) and shows homology to vinculin. The M_r 84,000 protein coimmunoprecipitated with E-cadherin has an electrophoretic mobility similar to that of plakoglobin (M_r 83,000), a cytoplasmic protein localized to different adherens junctions in epithelial cells (Cowin *et al.*, 1986). The M_r 102,000 and 92,000 E-cadherin-associated proteins also do not react with plakoglobin antibodies. However, we have detected plakoglobin (M_r 84,000) by Western blotting in stringently washed E-cadherin immunoprecipitates, suggesting that plakoglobin is part of the E-cadherin/cytoplasmic protein complex (Pipenhagen and Nelson, 1993; McCrea and Gumbiner, 1991a). Interestingly, indirect immunofluorescence of MDCK cells at short times after the induction of cell–cell contact shows distinct temporal differences in the spatial reorganization of E-cadherin and plakoglobin to sites of cell–cell contact, indicating that formation of a complex between these proteins occurs after cell–cell contact (Pipenhagen and Nelson, 1993). Limited peptide and complete nucleotide sequence analysis of β-catenin (M_r 92,000) has shown that this E-cadherin-associated protein is highly homologous to *armadillo* (McCrae *et al.*, 1991), a segment polarity gene in *Drosophila* which also shows $\sim 63\%$ homology to plakoglobin (Pfeifer and Wieschaus, 1990). This is particularly interesting in light of our observation that mono- and polyclonal plakoglobin antibodies raised against sodium dodecyl sulfate (SDS)-denatured plakoglobin do not react with any of the catenins, but do detect plakoglobin in MDCK cells and cell extracts. This suggests that there is a family of *armadillo*/plakoglobin-related proteins in these cells. The functions of these proteins in the assembly and organization of cell–cell junctions are unknown at present.

We have taken another experimental approach to analyzing interactions between E-cadherin and cytoplasmic (cytoskeleton) proteins in which native protein complexes between E-cadherin and cytoplasmic proteins are isolated by fractionating MDCK cell extracts in sucrose gradients and nondenaturing polyacrylamide gels (Nelson and Hammerton, 1989; Nelson *et al.*, 1990b). We have shown that ~30% of uvomorulin cosediments in sucrose gradients with the membrane-cytoskeletal proteins, ankyrin and fodrin, at ~10.5S (Nelson *et al.*, 1990b). These fractions are well separated in the gradient from the majority of proteins extracted from MDCK cells. The major peak of E-cadherin sediments at 7.5S but does not contain ankyrin or fodrin; we detected the presence of the M_r 102,000, 92,000, and 84,000 proteins in both fractions of E-cadherin in the sucrose gradient. Further separation of the 10.5S fraction in nondenaturing polyacrylamide gels revealed a complex of E-cadherin, ankyrin, and fodrin. Significantly, the 7.5S peak of E-cadherin migrated faster in the nondenaturing gel than the 10.5S complex. Taken together, these results indicate that there is more than one type of cadherin–cytoplasmic protein complex in these cells and that the membrane cytoskeleton is linked either directly or indirectly to a fraction of E-cadherin (Nelson *et al.*, 1990b).

V. ROLE OF CADHERINS IN CELL SURFACE REMODELING OF MEMBRANE PROTEINS

Our previous studies showed that induction of cell–cell contact in MDCK cells results in assembly of the membrane cytoskeleton at sites of cell–cell contact and the concomitant reorganization of Na,K-ATPase to those sites (Nelson *et al.*, 1990a). We have hypothesized that the membrane cytoskeleton is directly involved in determining the membrane distribution of Na,K-ATPase in these cells. Na,K-ATPase binds with high affinity to ankyrin (Nelson and Veschnock, 1987; Morrow *et al.*, 1989), and a protein complex comprising ankyrin, fodrin, and Na,K-ATPase has been isolated under conditions similar to those used to isolate the ankyrin–fodrin–E-cadherin complex described above (Nelson and Hammerton, 1989). We have suggested that induction of E-cadherin-mediated cell–cell contacts results in the redistribution of Na,K-ATPase to the contact zone of cells through a common linkage between these proteins to the membrane cytoskeleton (Nelson *et al.*, 1990a). To test this hypothesis directly, we analyzed the distribution of these proteins in fibroblasts (L-cells) that expressed no E-cadherin (control cells), or that were transfected with either full-length E-cadherin or E-cadherin with 37 or 72 amino acid deletions to the C-terminal portion of the cytoplasmic domain (McNeill *et al.*, 1990).

Control L-cells express Na,K-ATPase and fodrin uniformly over the surface even under confluent culture conditions that result in extensive, but nonspecific, cell–cell contacts. Significantly, in the presence of full-length E-cadherin, Na,K-ATPase and fodrin colocalize with E-cadherin at sites of cell–cell contact in a pattern similar to that in polarized MDCK cells. Deletions in the cytoplasmic domain of E-cadherin, however, result in only uniform distributions of Na,K-ATPase and fodrin over the cells, similar to those in cells that do not express E-cadherin. We interpret this result to show that deletions in the cytoplasmic domain of E-cadherin result in loss of binding to the membrane cytoskeleton and, hence, the loss of linkage between E-cadherin and Na,K-ATPase (McNeill *et al.*, 1990). Significantly, none of the catenins (Nagafuchi and Takeichi, 1988; Ozawa *et al.*, 1989,1990a) or membrane-cytoskeleton proteins (H. McNeill and W. J. Nelson, unpublished observations) are coimmunoprecipitated or cosediment with mutant cadherin containing deletions of the C-terminal 72 amino acids, indicating that these cytoplasmic proteins are normally bound to this domain; however, the binding site of E-cadherin for these proteins has not been determined.

We performed a series of controls to determine whether the induction of the reorganization of Na,K-ATPase and the membrane cytoskeleton by expression of E-cadherin in fibroblasts was common to other proteins (McNeill, *et al.*, 1990). We investigated two classes of proteins: major histocompatability (MHC) antigen and proteins that bound the lectin, wheat germ agglutinin (WGA). MHC molecules were located uniformly on the surface of fibroblasts and were absent from sites of cell–cell contact regardless of whether E-cadherin was expressed in the cells or not. In contrast, we found that a subpopulation of WGA-binding proteins were localized to sites of cell–cell contact; however, this distribution did not change in the presence or absence of E-cadherin expression. Thus, the redistribution of Na,K-ATPase and the membrane cytoskeleton in response to the expression of E-cadherin appears to be specific.

VI. REGULATION OF Na,K-ATPASE DISTRIBUTION DURING FORMATION OF POLARIZED EPITHELIAL CELLS

The studies described above provide qualitative information on the roles of E-cadherin and the membrane cytoskeleton in the development of cell surface polarity of Na,K-ATPase. More recently we have sought a quantitative approach to determine the intracellular sorting pathway of Na,K-ATPase and the fate of the protein on arrival at different membrane domains. A logical prediction from our previous studies is that Na,K-ATPase should become stabilized in the basal-lateral membrane in association with the membrane cytoskeleton. Therefore, we determined the delivery and fate of Na,K-ATPase at different cell surface domains in MDCK cells during the development of cell surface polarity (Hammerton *et al.*, 1991). Confluent monolayers of cells were established on permeable Transwell filter inserts; these inserts allow direct and independent access to either the apical or the basal-lateral cell surfaces which can then be labeled with biotinylated cross-linking reagents. The introduction of biotinylated cross-linking reagents, e.g., sulfo-NHS-biotin, to label cell surface proteins in monolayers of polarized epithelial cells has been an important advance for the quantitative analysis of protein distributions in these cells (Le Bivic *et al.*, 1990,1991; Sargiacomo *et al.*, 1989).

We established confluent monolayers of MDCK cells on filters, induced E-cadherin-dependent cell–cell contact, and then analyzed the development of cell surface polarity. Cell surface distribution of Na,K-ATPase was determined qualitatively by laser-scanning confocal microscopy and quantitatively by cell surface biotinylation. Following cell surface biotinylation, Na,K-ATPase subunits were immunoprecipitated from cell extracts with specific antibodies, separated by SDS–polyacrylamide gel electrophoresis (PAGE), and transferred to nitrocellulose and the biotinylated Na,K-ATPase was detected with [125I] streptavidin (Hammerton *et al.*, 1991). Our results showed that the generation of cell surface polarity of the Na,K-ATPase after induction of cell–cell contact was relatively slow (see below), requiring ~72 hr, by which time >90% of the Na,K-ATPase was located on the basal-lateral membrane. During the development of polarity, we found that Na,K-ATPase was transiently expressed on the apical membrane.

Previous studies of MDCK cells have indicated that cell surface polarity is maintained by vectorial delivery of proteins from the Golgi complex to the correct membrane (reviewed in Rodriguez-Boulan and Nelson, 1989; Simons and Fuller, 1985). We, therefore, investigated the delivery of newly synthesized Na,K-ATPase to the cell surface during the development of cell surface polarity (Hammerton *et al.*, 1991). Cells on pairs of filters were metabolically labeled and biotinylated on either the apical or the basal-lateral membrane of pairs of filters, and Na,K-ATPase subunits were immunoprecipitated with specific antibodies from cell extracts. The immunoprecipitates were dissociated in SDS or low pH, and biotinylated proteins recovered subsequently by precipitation with avidin agarose;

proteins were detected by SDS–PAGE and fluorography. Results showed clearly that Na,K-ATPase subunits were not targeted specifically to the basal-lateral membrane at up to 6–7 days after the generation of cell surface polarity at steady state (Hammerton *et al.*, 1991); instead, Na,K-ATPase was delivered approximately equally to both the apical and the basal-lateral membrane domains.

How was steady-state polarity of Na,K-ATPase achieved when newly synthesized protein was being delivered to both the apical and the basal-lateral membranes? In a pulse-chase experiment, together with the detection of protein on the membrane by biotinylated cross-linking as described above, we determined the half-life (also termed the residence time) of newly synthesized Na,K-ATPase on the apical and basal-lateral membranes, respectively (Hammerton *et al.*, 1991). We found that Na,K-ATPase delivered to the basal-lateral membrane was stabilized ($t_{0.5} > 36$ hr), while Na,K-ATPase delivered to the apical membrane was rapidly removed ($t_{0.5}$ 1–2 hr). Thus, cell surface polarity of Na,K-ATPase appears to be regulated by differences in the stability of the protein in the apical and basal-lateral membranes (Fig. 3). Furthermore, stabilization of Na,K-

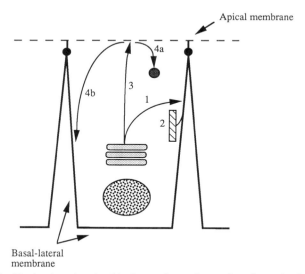

FIGURE 3   Mechanisms involved in the sorting and retention of proteins in the development of cell surface polarity of a basal-lateral membrane protein; for details see the text. 1, basal-lateral delivery; 2, basal-lateral retention; 3, apical delivery; 4, internalization from apical membrane; 4a, degradation in lysosomes; 4b, transcytosis.

ATPase in the basal-lateral membrane correlates with binding of the protein to the membrane cytoskeleton (Nelson *et al.,* 1990a).

In addition, we analyzed the distribution of active Na,K-ATPase by measuring ouabain-binding sites on the plasma membrane at different times after the induction of cell–cell contact (Hammerton *et al.,* 1991); ouabain is a specific inhibitor of active Na,K-ATPase (Jorgensen, 1986). We found that ouabain-binding sites were lost from the apical membrane within 12–24 hr after induction of cell–cell contact. However, at these times there were still substantial quantities of Na,K-ATPase located on the apical membrane domain. This result strongly indicates that Na,K-ATPase delivered to the apical membrane was not active and thus did not bind ouabain; this result may also explain why a previous investigation that used ouabain to detect Na,K-ATPase did not find Na,K-ATPase on the apical membrane of these cells (Caplan *et al.,* 1986).

At present the mechanism(s) involved in the activation of Na,K-ATPase in the apical membrane are unknown. However, the apical membrane of renal epithelial cells contains a high concentration of glycosphingolipids compared to the basal-lateral membrane (van Meer and Simons, 1988), and previous studies indicate that Na,K-ATPase activity may be modulated by the lipid environment of the membrane bilayer (Kimelberg and Papahadjo-poulos, 1972,1974; Palatini *et al.,* 1977). Thus, it is possible that the lipid environment of the apical membrane may inactivate Na,K-ATPase. To test this possibility, we have purified Na,K-ATPase to homogeneity and reconstituted the enzyme into proteoliposomes containing an increasing proportion of glycosphingolipids. Our preliminary results indicate that increased glycosphingolipid content in reconstituted proteoliposomes results in a marked decrease in Na,K-ATPase activity as measured by the rate of ATP hydrolysis (Mays and Nelson, unpublished observation). Since the apical membrane of normal polarized renal epithelial cells contains glycosphingolipids (van Meer and Simons, 1988), it is possible that glycosphingolipids directly affect Na,K-ATPase activity *in vivo.*

VII. REGULATION OF MEMBRANE-CYTOSKELETON ASSEMBLY

The results describes above indicate that assembly of the membrane cytoskeleton plays an important role in retention of Na,K-ATPase in the basal-lateral membrane of these cells and, therefore, in determining the membrane distribution of this important protein. At present, however, we do not know the mechanism(s) involved in inducing assembly of the membrane cytoskeleton at cell–cell contacts. The membrane cytoskeleton comprises several well-characterized proteins including ankyrin, fodrin (spectrin), adducin, protein 4.1, and actin (reviewed in Bennett, 1990). Adducin, protein 4.1, and actin associate with cell–cell contacts in MDCK cells (Geiger, 1989; Kaiser *et al.,* 1989; J. A. Marrs, K. A. Krzeminski, and W. J. Nelson, unpublished observations) and are thought to function in the interaction of spectrin and actin filaments (Bennett, 1989,1990; Nelson, 1989; Nelson and Lazarides, 1984). It is noteworthy that protein interactions in the membrane cytoskeleton are regulated by Ca^{2+} and phosphorylation (reviewed in Bennett, 1989,1990); for instance, phosphorylation of ankyrin lowers its affinity for spectrin tetramers (Lu *et al.,* 1985; Lu and Tao, 1986), phosphorylation of protein 4.1 decreases its affinity for spectrin and its ability to mediate spectrin–actin interactions (Eder *et al.,* 1986; Ling *et al.,* 1988), and adducin bundles actin filaments and is a substrate for phosphorylation by protein kinase C (Gardner and Bennett, 1987; Ling *et al.,* 1988; Mische *et al.,* 1987; Palfrey and Waseem, 1985). These posttranslational modifications may play an important role in modulating membrane-cytoskeleton assembly at sites of cell–cell adhesion.

VIII. REGULATION OF DEVELOPMENT OF CELL SURFACE POLARITY OF OTHER MEMBRANE PROTEINS

Studies of the remodeling of Na,K-ATPase distribution as a function of time after cell–cell contact indicate that the development of cell surface polarity of this protein is determined by retention of protein in the basal-lateral membrane, rather than direct delivery of protein from the Golgi complex to the basal-lateral membrane as is thought to occur for the majority of proteins studied in these cells (Simons and Fuller, 1985; Rodriguez-Boulan and Nelson, 1989). Is this mechanism important in the development of cell surface polarity of other membrane proteins in these cells? To address this question, we have analyzed the development of cell surface polarity of two other basal-lateral membrane proteins, E-cadherin and desmoglein-1 (Dg-1, a desmosome-specific cell adhesion protein; Koch *et al.,* 1990), and two apical proteins, gp 135 and the secreted protein gp81 (Wollner *et al.,* 1992). We found that the distributions of E-cadherin and Dg-1 became restricted to the basal-lateral membrane domain within 8 hr of cell–cell contact. During this time, however, 60–80% of the newly synthesized Dg-1 and E-cadherin was delivered directly to the forming apical membrane and then rapidly removed ($t_{0.5} < 2$ hr), while the remainder that was delivered to the basal-lateral membrane had a longer residence time ($t_{0.5} > 24$ hrs). Direct delivery of >95% of these proteins from the Golgi complex to the basal-lateral membrane occurred >48 hr later. In contrast, we found that two apical proteins were efficiently delivered and restricted to the apical cell surface within 2 hr following cell–cell contact (Wollner *et al.*, 1992). These results indicate that selective retention in the plasma membrane may be an important mechanism for developing and, in some cases, maintaining the cell surface distribution of membrane proteins in this cell type.

The question arose: When is direct delivery of proteins from the Golgi complex to the apical or basal-lateral membrane established in these cells? We sought to determine whether protein sorting occurred at the time of induction of cell–cell contact by analyzing the direction of secretion of gp81, the major apically secreted protein in polarized MDCK cells (Gottlieb *et al.*, 1986; Kondor-Koch *et al.*, 1985, Urban *et al.*, 1987). We showed that >85% of newly synthesized gp81 is secreted directly into the apical medium within 15 min of the induction of cell–cell contact (Wollner *et al.*, 1992). This result indicates that shortly after cell–cell contact intracellular sorting of gp81 occurs prior to delivery to the cell surface. Significantly, at the earliest time that delivery of membrane proteins to the cell surface could be analyzed (2 hr after cell–cell contact), we also detected delivery of >70% of both apical and basal-lateral membrane proteins to the apical membrane (Wollner *et al.*, 1992).

There are at least two possible explanations of this result (Wollner *et al.*, 1992). (i) There is little or no sorting of apical and basal-lateral proteins in the Golgi complex at this time and >70% of all vesicles are delivered directly to the forming apical membrane; this implies that time is required to establish the putative sorting machinery in the Golgi complex. (ii) Sorting of apical and basal-lateral membrane proteins in the Golgi complex occurs constitutively in these cells regardless of the state of cell surface polarity but, initially, delivery of both populations of vesicles is directed to the apical membrane; this implies that vesicle delivery to the apical surface is regulated at this time. (In either case, it is possible that Na,K-ATPase has a relatively weak signal for sorting into the basal-lateral pathway in these cells.) At present, we are not able to distinguish unequivocally between these possibilities, although the finding that gp81 is delivered directly to the apical membrane indicates that sorting of apical and basal-lateral proteins may occur prior to the development of cell surface polarity; more recently, we have found that gp81 is directly secreted from the apical membrane in cells grown in the absence of cell–cell contact (Wollner and Nelson, unpublished observation).

That the majority of newly synthesized Dg-1, E-cadherin, and gp135/170 are initially delivered to the apical membrane indicates the presence of a facilitated pathway of delivery of vesicles from the Golgi complex to the apical membrane at this time. Previous studies have shown that microtubules facilitate the delivery of vesicles to the apical membrane in fully polarized MDCK cells (Achler *et al.*, 1989; Breitfeld *et al.*, 1990; Gilbert *et al.*, 1991; Hunziker *et al.*, 1990; Matter *et al.*, 1990; Parczyk *et al.*, 1989; Rindler *et al.*, 1987; Salas *et al.*, 1986; Van Zeijl and Matlin, 1990). Depolymerization of microtubules with drugs causes a decrease in the transport of proteins to the apical membrane and some missorting of proteins to the basal-lateral membrane (Gilbert *et al.*, 1991; Parczyk *et al.*, 1989; Van Zeijl and Matlin, 1990). The presence of a facilitated pathway of vesicle delivery from the Golgi complex to the apical cell surface in single MDCK cells and in cells shortly after the induction of cell–cell contact may provide an explanation for the polarity of apical proteins in these cells. It will be important to determine whether the disruption of microtubules affects the delivery of vesicles to the apical surface during the development of cell surface polarity following the induction of cell–cell contact in MDCK cells.

Our studies indicate that direct delivery of E-cadherin and Dg-1 to the basal-lateral membrane domain does not occur until >48 hr after the induction of cell–cell contact (Wollner *et al.*, 1992). This may reflect the time required to reorganize the part of the cytoskeleton that allows delivery of vesicles to the basal-lateral membrane, or the time required for these vesicles to achieve the capacity to recognize and fuse with the basal-lateral plasma membrane. Docking proteins on the basal-lateral membrane may be involved as a recognition system necessary for basal-lateral vesicle fusion with that plasma membrane domain and may not be in position in sufficient quantity prior to this time to accept vesicles being delivered from the Golgi complex.

IX. MECHANISMS OF INDUCTION OF CELL MORPHOGENESIS BY CADHERINS

In summary, there is a strong evidence that cadherins mediate cell–cell recognition and adhesion and that direct linkage between cadherins and cytoplasmic (cytoskeletal) proteins is important in cell adhesion and cellular morphogenesis. However, the mechanism(s) involved in transducing extracellular cell–cell adhesion, through cadherins, into the reorganization of cell structure and function is unknown. Several levels of complexity are possible. The mechanism could involve increased local concentrations of cadherins at sites of cell–cell contact that cause multivalent stabilization of weak interactions with cytoplasmic proteins through the cytoskeleton. Several examples of this type of mechanism have been found, including the activation of complement by the binding of Clq to cell surface antigen–IgG complexes and cellular degranulation triggered by extracellular cross-linking of IgE–receptor complexes (Kane *et al.*, 1988). Signal transduction could (also) be involved in which occupancy of cadherin results in the activation of a classical signal transduction response through second messengers (Ca^{2+}, inositol phosphates, protein kinase C; reviewed in Berridge and Irvine, 1989; Freissmuth *et al.*, 1989; Kikkawa *et al.*, 1989; Rasmussen and Rasmussen, 1990). Evidence that second messengers may be involved has been reported by Schuch *et al.* (1989), who demonstrated that addition of antibodies against the extracellular domain of the cell adhesion proteins N-CAM or L1 to PC12 cells induced an increase in intracellular Ca^{2+} levels and a reduction in inositol phosphates and intracellular pH. These changes were inhibited by pertussin toxin, indicating involvement of a G-protein in the transduction process (Schuch *et al.*, 1989). At present, we do not know whether the induction of cell surface remodeling that is induced by the interaction of cells through E-cadherin is regulated by a signal transduction pathway involving second messengers or by the process of patching cell surface proteins through interactions with the membrane cytoskeleton. An understanding of this process is likely to provide an important advancement in our knowledge of how extracellular activation of cell adhesion proteins is transduced into the structural and functional remodeling of cells.

Acknowledgments

Work from the author's laboratory was supported by grants from the NIH and the March of Dimes and by an Established Investigator Award.

References

Achler, C., Filmer, D., Merte, C., and Drenckhahn, D. (1989). Role of microtubules in polarized delivery of apical membrane proteins to the brush border of the intestinal epithelium. *J. Cell Biol.* **109**, 179–189.

Behrens, J., Birchmeier, W., Goodman, S. L., and Imhof, B. A. (1985). Dissociation of Madin–Darby canine kidney epithelial cells by the monoclonal antibody anti-arc-1: mechanistic aspects and identification of the antigen as a component related to uvomorulin. *J. Cell Biol.* **101**, 1307–1315.

Behrens, J., Mareel, M. M., Van, R. F., and Mirchmeier, W. (1989). Dissecting tumor cell invasion: epithelial cells acquire invasive properties after the loss of uvomorulin-mediated cell-cell adhesion. *J. Cell Biol.* **108**, 2435–2447.

Bennett, V. (1989). The spectrin-actin junction of erythrocyte membrane skeletons. *Biochim. Biophys. Acta* **988**, 107–121.

Bennett, V. (1990). Spectrin-based membrane skeleton: A multipotential adaptor between plasma membrane and cytoplasm. *Physiol. Rev.* **70**, 1029–1065.

Berridge, M. J., and Irvine, R. F. (1989). Inositol phosphates and cell signalling. *Nature (London)* **341**, 197–205.

Blaschuk, O. W., Pouliot, Y., and Holland, P. C. (1990). Identification of a conserved region common to cadherins and influenza strain A hemmagglutinins. *J. Dev. Biol.* **211**, 679–682.

Breitfield, P. P., McKinnon, W. C., and Mostov, K. E. (1990). Effect of nocodazole on vesicular traffic to the apical and basolateral surfaces of polarized MDCK cells. *J. Cell Biol.* **111,** 2365–2373.

Caplan, M. J., Anderson, H. C., Palade, G. E., and Jamieson, J. D. (1986). Intracellular sorting and polarized cell surface delivery of Na^+, K^+-ATPase, an endogenous component of MDCK cell basolateral plasma membranes. *Cell* **46,** 623–631.

Cowin, P., Kapprell, H.-P., Franke, W. W., Tamkun, J., and Hynes, R. O. (1986). Plakolobin: a protein common to different kinds of intercellular adhering junctions. *Cell* **46,** 1063–1073.

Cunningham, B. A., Hemperly, J. J., Murray, B. A., Prediger, E. A., Brackenbury, R., and Edelman, G. M. (1987). Neural cell adhesion molecule: Structure, immunoglobulin-like domains, cell surface modulation, and alternative RNA splicing. *Science* **236,** 799–806.

Edelman, G. M., and Crossin, K. L. (1991). Cell adhesion molecules: implications for a molecular histology. *Annu. Rev. Biochem.* **60,** 155–190.

Eder, P., Soong, C., and Tao, M. (1986). Phosphorylation reduces the affinity of protein 4.1 for spectrin. *Biochemistry* **25,** 1764–1770.

Freissmuth, M., Casey, P. J., and Gilman, A. G. (1989). G proteins control diverse pathways of transmembrane signaling. *FASEB J.* **3,** 2125–2131.

Gallin, W. J., Chuong, C. M., Finkel, L. H., and Edelman, G. M. (1986). Antibodies to L-CAM perturb inductive interactions and alter feather pattern and structure. *Proc. Natl. Acad. Sci. U.S.A.* **83,** 8235–8239.

Gardner, K., and Bennett, V. (1987). Modulation of spectrin–actin assembly by erythrocyte adducin. *Nature (London)* **328,** 359–362.

Geiger, B. (1989). Cytoskeleton-associated cell contacts. *Curr. Opinion Cell Biol.* **1,** 103–109.

Gilbert, T., Lebivic, A., Quaroni, A., and Rodriguez-Boulan, E. (1991). Microtubular organization and its involvement in the biogenetic pathways of plasma membrane proteins in CaCo-2 intestinal epithelial cells. *J. Cell Biol.* **113,** 275–284.

Gottlieb, T. A., Beaudry, G., Rizzolo, L., Colman, A., Rindler, M., Adesnik, M., and Sabatini, D. D. (1986). Secretion of endogenous and exogenous proteins from polarized MDCK cell monolayers. *Proc. Natl. Acad. Sci. U.S.A.* **83,** 2100–2104.

Gumbiner, B., Stevenson, B., and Grimaldi, A. (1988). The role of the cell adhesion molecule uvomorulin in the formation and maintenance of the epithelial junctional complex. *J. Cell Biol.* **107,** 1575–1587.

Hammerton, R. W., Krzeminski, K. A., Mays, R. W., Ryan, T. A., Wollner, D. A., and Nelson, W. J. (1991). Mechanism for regulating cell surface distribution of Na^+,K^+-ATPase in polarized epithelial cells. *Science* **254,** 847–850.

Hatta, K., Okada, T. S., and Takeichi, M. (1985). A monoclonal antibody disrupting calcium-dependent cell–cell adhesion of brain tissues: possible role of its target antigen in animal pattern formation. *Proc. Natl. Acad. Sci. U.S.A.* **82,** 2789–2793.

Hatta, K., Nose, A., Nagafuchi, A., and Takeichi, M. (1988). Cloning and expression of cDNA encoding a neural calcium-dependent cell adhesion molecule: its identity in the cadherin gene family. *J. Cell Biol.* **106,** 873–881.

Hirai, Y., Nose, A., Kobayashi, S., and Takeichi, M. (1989). Expression and role of E- and P-cadherin adhesion molecules in embryonic histogenesis II. Skin morphogenesis. *Development* **105,** 271–277.

Hirano, S., Nose, A., Hatta, K., Kawakami, A., and Takeichi, M. (1987). Calcium-dependent cell–cell adhesion molecules (cadherins): subclass specificities and possible involvement of actin bundles. *J. Cell Biol.* **105,** 2501–2510.

Hoffman, S., and Edelman, G. M. (1983). Kinetics of homophilic binding by embryonic and adult forms of the neural cell adhesion molecule. *Proc. Natl. Acad. Sci. U.S.A.* **80**, 5762–5766.

Hunziker, W., Vale, P., and Mellman, I. (1990). Differential microtubule requirements for transcytosis in MDCK cells. *EMBO J.* **9**, 3515–3525.

Hyafil, F., Babinet, C., and Jacob, F. (1981). Cell–cell interactions in early embryogenesis: a molecular approach to the role of calcium. *Cell* **26**, 447–454.

Jorgensen, P. L. (1986). Mechanism of the Na^+,K^+-ATPase pump: protein structure and conformations of the pure Na^+,K^+-ATPase. *Biochim. Biophys. Acta* **694**, 27–68.

Kaiser, H. W., O'Keefe, E., and Bennett, V. (1989). Adducin: Ca^{++}-dependent association with sites of cell–cell contact. *J. Cell Biol.* **109**, 557–569.

Kane, P. M., Holowka, D., and Baird, B. (1988). Cross-linking of IgE-receptor complexes by rigid bivalent antigens >200A in length triggers cellular degranulation. *J. Cell Biol.* **107**, 969–980.

Kikkawa, U., Kishimoto, A., and Nishizuka, Y. (1989). The protein kinase C family: Heterogeneity and its implications. *Annu. Rev. Biochem.* **58**, 31–44.

Kimelberg, H. K., and Papahadjopoulos, D. (1972). Phospholipid requirements for Na/K-ATPase activity: head-group specificity and fatty acid fluidity. *Biochim. Biophys. Acta* **282**, 277–292.

Kimelberg, H. K., and Papahadjopoulos, D. (1974). Effects of phospholipid acyl chain fluidity, phase transitions, and cholesterol on Na/K-ATPase-stimulated adenosine triphosphatase. *J. Biol. Chem.* **249**, 1071–1080.

Koch, P. J., Walsh, M. J., Schmelz, M., Goldschmidt, M. D., Zimbelmann, R., and Franke, W. W. (1990). Identification of desmoglein, a constitutive desmosomal glycoprotein, as a member of the cadherin family of cell adhesion molecules. *Eur. J. Cell Biol.* **53**, 1–12.

Kondor-Koch, C., Bravo, R., Fuller, S. D., Cutler, D., and Garoff, H. (1985). Exocytotic pathways exist to both the apical and the basolateral cell surface of the polarized epithelial cell MDCK. *Cell* **43**, 297–306.

Le Bivic, A., Sambuy, Y., Mostov, K., and Rodriguez-Boulan, E. (1990). Vectorial targeting of an endogenous apical membrane sialoglycoprotein and uvomorulin in MDCK cells. *J. Cell Biol.* **110**, 1533–1539.

Le Bivic, A., Quaroni, A., Nichols, B., and Rodriguez-Boulan, E. (1991). Biogenetic pathways of plasma membrane proteins in Caco-2, a human intestinal epithelial cell line. *J. Cell Biol.* **111**, 1351–1362.

Ling, E., Danilov, Y., and Cohen, C. (1988). Modulation of red cell band 4.1 function by AMP-dependent kinase and protein kinase C phosphorylation. *J. Biol. Chem.* **263**, 2209–2216.

Lu, P. W., and Tao, M. (1986). Phosphorylation of protein tyrosine by human erythrocyte casein kinase A. *Biochem. Biophys. Res. Commun.* **139**, 855–860.

Lu, P. W., Soong, C. J., and Tao, M. (1985). Phosphorylation of ankyrin decreases its affinity for spectrin tetramer. *J. Biol. Chem.* **260**, 14958–14964.

Matter, K., Bucher, K., and Hauri, H.-P. (1990). Microtubule perturbation retards both the direct and the indirect apical pathway but does not affect sorting of plasma membrane proteins in intestinal epithelial cells. (Caco-2). *EMBO J.* **9**, 3163–3170.

McCrae, P. D., and Gumbiner, B. (1991a). Purification of a 92kDa cytoplasmic protein tightly associated with the cell–cell adhesion molecule E-cadherin (Uvomorulin). *J. Biol. Chem.* **266**, 4514–4520.

McCrea, P. D., Turck, C. W., and Gumbiner, B. (1991b). A homolog of the armadillo protein in *Drosophila* (plakoglobin) associated with E-cadherin. *Science* **254**, 1350–1361.

McNeill, H., Ozawa, M., Kemler, R., and Nelson, W. J. (1990). Novel function of the cell adhesion molecule uvomorulin as an inducer of cell surface polarity. *Cell* **62,** 309–316.

Mege, R. M., Matsuzaki, F., Gallin, W. J., Goldberg, J. I., Cunningham, B. A., and Edelman, G. M. (1988). Construction of epitheliod sheets by transfection of mouse sarcoma cells with cDNAs for chicken cell adhesion molecules. *Proc. Natl. Acad. Sci. U.S.A.* **85,** 7274–7278.

Mische, S., Mooseker, M. S., and Morrow, J. S. (1987). Erythrocyte adducin: A calmodulin-regulated actin-bundling protein that stimulates spectrin–actin binding. *J. Cell Biol.* **105,** 2837–2849.

Morrow, J. S., Cianci, C. D., Ardito, T., Mann, A. S., and Kashgarian, M. (1989). Ankyrin links fodrin to the alpha subunit of Na^+,K^+,ATPase in Madin–Darby canine kidney cells and in intact renal tubule cells. *J. Cell Biol.* **108,** 455–465.

Nagafuchi, A., and Takeichi, M. (1988). Cell binding function of E-cadherin is regulated by the cytoplasmic domain. *Embo J.* **7,** 3679–3684.

Nagafuchi, A., Shirayoshi, Y., Okazaki, K., Yasuda, K., and Takeichi, M. (1987). Transformation of cell adhesion properties by exogenously introduced E-cadherin cDNA. *Nature (London)* **329,** 340–343.

Nagafuchi, A., Takeichi, M., and Tsukita, S. (1991). The 102-kDa cadherin-associated protein: Similarity to vinculin and posttranscriptional regulation of expression. *Cell* **65,** 849–857.

Nelson, W. J. (1989). Development and maintenance of epithelial polarity: A role for the submembranous cytoskeleton. *In* "Modern Cell Biology: Functional Epithelial Cells in Culture" (K. Matlin and J. R. Valentich, eds.), Vol. **8,** pp. 3–42. A. R. Liss, New York.

Nelson, W. J., and Hammerton, R. W. (1989). A membrane-cytoskeletal complex containing Na^+,K^+-ATPase, ankyrin, and fodrin in Madin–Darby canine kidney (MDCK) cells: Implications for the biogenesis of epithelial cell polarity. *J. Cell Biol.* **108,** 893–902.

Nelson, W. J., and Lazarides, E. (1984). Assembly and establishment of membrane–cytoskeleton domains during differentiation: spectrin as a model system. *In* "Cell Membranes: Methods and Reviews" (E. Elson, W. Frazier, and L. Glaser, eds.), Vol. 2, pp. 219–246. Plenum, New York.

Nelson, W. J., and Veshnock, P. J. (1987). Ankyrin binding to Na^+,K^+-ATPase and implications for the organization of membrane domains in polarized cells. *Nature (London)* **328,** 533–536.

Nelson, W. J., Hammerton, R. W., Wang, A. Z., and Shore, E. M. (1990a). Involvement of the membrane–cytoskeleton in the development of epithelial cell polarity. *Semin. Cell Biol.* **1,** 359–371.

Nelson, W. J., Shore, E. M., Wang, A. Z., and Hammerton, R. W. (1990b). Identification of a membrane-cytoskeletal complex containing the cell adhesion molecule uvomorulin (E-cadherin), ankyrin, and fodrin in Madin–Darby canine kidney epithelial cells. *J. Cell Biol.* **110,** 349–357.

Ozawa, M., Baribault, H., and Kemler, R. (1989). The cytoplasmic domain of the cell adhesion molecule uvomorulin associates with three independent proteins structurally related in different species. *EMBO J.* **8,** 1711–1717.

Ozawa, M., Engel, J., and Kemler, R. (1990a). Single amino acid substitutions in one Ca2 + binding site of uvomorulin abolish the adhesive function. *Cell* **63,** 1033–1038.

Ozawa, M., Ringwald, M., and Kemler, R. (1990b). Uvomorulin–catenin complex formation is regulated by a specific domain in the cytoplasmic region of the cell adhesion molecule. *Proc. Natl. Acad. Sci. U.S.A.* **87,** 4246–4250.

Palatini, P., Dabbeni-Sala, F., Pitotti, A., Bruni, A., and Mandersloot, J. C. (1977). Activation of Na/K-dependent ATPase by lipid vesicles of negative phospholipids. *Biochim. Biophys. Acta* **466**, 1–9.

Palfrey, H. C., and Waseem, A. (1985). Protein kinase C in the human erythrocyte. Translocation to the plasma membrane and phosphorylation of bands 4.1 and 4.9 and other membrane proteins. *J. Biol. Chem.* **260**, 16021–16029.

Parczyk, K., Hasse, W., and Kondor, K. C. (1989). Microtubules are involved in the secretion of proteins at the apical cell surface of the polarized epithelial cell, MDCK. *J. Biol. Chem.* **264**, 16837–16846.

Pfeifer, M., and Wieschaus, E. (1990). The segment polarity gene armadillo encodes a functionally modular protein that is the *Drosophila* homolog of human plakoglobin. *Cell* **63**, 1167–1178.

Piepenhagen, P. A., and Nelson, W. J. (1993). Defining E-cadherin-associated protein complexes in epithelial cells: plakoglobin, β- and γ-catenin are distinct components. *J. Cell Sci.,* **104**, 751–762.

Rasmussen, H., and Rasmusen, J. E. (1990). Calcium as intracellular messenger: from simplicity to complexity. *Curr. Top. Cell. Regul.* **31**, 1–82.

Rindler, M. J., Ivanov, I. E., and Sabatini, D. D. (1987). Microtubule-acting drugs lead to the nonpolarized delivery of the influenza hemagglutinin to the cell surface of polarized Madin–Darby canine kidney cells. *J. Cell Biol.* **104**, 231–241.

Ringwald, M., Schuh, R., Vestweber, D., Eistetter, H., Lottspeich, F., Engel, J., Dolz, R., Jahnig, F., Epplen, J., Mayer, S., Muller, C., and Kemler, R. (1987). The structure of the cell adhesion molecule uvomorulin. Insights into the molecular mechanisms of Ca^{2+}-dependent cell adhesion. *EMBO J.* **6**, 3647–3653.

Rodriguez-Boulan, E., and Nelson, W. J. (1989). Morphogenesis of the polarized epithelial cell phenotype. *Science* **245**, 718–725.

Salas, P. J., Misek, D. E., Vega, S. D., Gundersen, D., Cereijido, M., and Rodriguez, B. E. (1986). Microtubules and actin filaments are not critically involved in the biogenesis of epithelial cell surface polarity. *J. Cell Biol.* **102**, 1853–1867.

Sargiacomo, M., Lisanti, M., Graeve, L., Le Bivic, A., and Rodriguez-Boulan, E. (1989). Integral and peripheral protein composition of the apical and basolateral membrane domains in MDCK cells. *J. Membr. Biol.* **107**, 277–286.

Schuch, U., Lohse, M. J., and Schachner, M. (1989). Neural cell adhesion molecules influence second messenger systems. *Neuron* **3**, 13–20.

Shimoyama, Y., Hirohashi, S., Hirano, S., Noguchi, M., Shimosato, Y., Takeichi, M., and Abe, O. (1989). Cadherin cell-adhesion molecules in human epithelial tissues and carcinomas. *Cancer Res.* **49**, 2128–2133.

Shore, E. M., and Nelson, W. J. (1991). Biosynthesis of the cell adhesion molecule uvomorulin (E-cadherin) in Madin–Darby canine kidney (MDCK) epithelial cells. *J. Biol. Chem.* **266**, 19672–19680.

Simons, K., and Fuller, S. D. (1985). Cell surface polarity in epithelia. *Annu. Rev. Cell Biol.* **1**, 243–288.

Springer, T. A. (1990). Adhesion molecules of the immune system. *Nature (London)* **346**, 425–433.

Takeichi, M. (1988). The cadherins: Cell–cell adhesion molecules controlling animal morphogenesis. *Development* **102**, 639–655.

Takeichi, M. (1990). Cadherins: A molecular family important in selective cell–cell adhesion. *Annu. Rev. Biochem.* **59**, 237–252.

Takeichi, M. (1991). Cadherin cell adhesion receptors as a morphogenetic regulator. *Science* **251**, 1451–1455.

Trinkaus, J. P. (1984). "Cells into Organs." Prentice-Hall, Englewood Cliffs, New Jersey.

Urban, J., Paraczyk, K., Leutz, A., Kayne, M., and Kondor-Koch, C. (1987). Constitutive apical secretion of an 80-kDa sulfated glycoprotein complex ion the polarized epithelial Madin–Darby canine kidney cell line. *J. Cell Biol.* **105,** 2735–2743.

van Meer, G., and Simons, K. (1988). Lipid sorting in epithelial cells. *Biochemistry* **27,** 6197–6202.

Van Zeijl, M. J. A. H., and Matlin, K. S. (1990). Microtubule perturbation inhibits intracellular transport of an apical membrane glycoprotein in a substrate-dependent manner in polarized Madin–Darby canine kidney epithelial cells. *Cell Regul.* **1,** 921–936.

Vestweber, D., and Kemler, R. (1984). Rabbit antiserum against a purified surface glycoprotein decompacts mouse preimplantation embryos and reacts with specific adult tissues. *Exp. Cell Res.* **152,** 169–178.

Vestweber, D., and Kemler, R. (1985). Identification of a putative cell adhesion domain of uvomorulin. *EMBO J.* **4,** 3393–3398.

Vestweber, D., Kemler, R., and Ekblom, P. (1985). Cell-adhesion molecule uvomorulin during kidney development. *Dev. Biol.* **112,** 213–221.

Vleminckx, K., Vakaet, L., Mareel, M., Fiers, W., and Van Roy, F. (1991). Genetic manipulation of E-cadherin expression by epithelial tumor cells reveals an invasion suppressor role. *Cell* **66,** 107–120.

Wollner, D. A., Krzeminski, K. A., and Nelson, W. J. (1992). Remodelling the cell surface distribution of membrane proteins during the development of epithelial cell polarity. *J. Cell Biol.* **116,** 889–899.

Note added in proof: Another clone of MDCK cells appears to sort Na,K-ATPase to the basal-lateral membrane (Gottardi, C. J., and Caplan, M. J. (1993). *Science* **260,** 552–554). Differences in sorting are not due to experimental differences (Siemess, K. A. *et al.* (1993). *Science* **260,** 554–556), but to differences in protein and lipid sorting pathways between the cell clones (Mays, R. W., van Meer, G., and Nelson, W. J., in preparation).

CHAPTER 8

Synthesis and Sorting of Ion Pumps in Polarized Cells

Cara J. Gottardi,* Grazia Pietrini,† Monica J. Shiel,* and Michael J. Caplan*

*Department of Cellular and Molecular Physiology, Yale University School of Medicine, New Haven, Connecticut 06510; and †CNR Department of Pharmacology, Chemotherapy and Medical Toxicology, University of Milan, 20129 Milan, Italy

I. INTRODUCTION

The word "polarity" is an example of a nontechnical term that has been appropriated by several different scientific disciplines, each of which has imbued its definition with field-specific connotations. There is some utility, therefore, in establishing the meaning of this word as well as in exploring its applications to the discussions contained within this volume. When invoked in one of its several cell biologic contexts, the term polarity relates to the division of a cell's plasma membrane into subdomains which manifest distinct biochemical compositions and functional properties. When this rather broad formulation is literally applied, evidence of polarity can be appreciated in a remarkably large menagerie of cell types. The most obvious exemplars of polarity are epithelia, whose plasmalemmas are divided into two morphologically distinguishable domains that are demarcated by intercellular junctional complexes and that confront separate body compartments (Caplan and Matlin, 1989; Rodriguez-Boulan and Nelson, 1989; Simons and Fuller, 1985; Simons and Wandinger-Ness, 1990). Equally dramatic polarity is exhibited by neurons, whose axonal and dendritic surfaces are equipped to subserve discrete steps in the process of information transmission (Dotti and Simons, 1990). Similarly, the surface membranes of mammalian sperm cells are divided into five sharply defined cylindrical zones characterized by the presence of unique protein antigens (Primakoff and Miles, 1983). Subtler manifestations of polarity can be found in the ruffled membranes of migrating fibroblasts or in the sealing zones of resorbing osteoclasts. Finally yeast as well as certain species of prokaryote manage to generate and maintain anisotropies in their plasma membranes (Nelson, 1992).

Given the enormous range of cell types and cellular phenomena which conform to our codification of polarity, one can be forgiven for wondering whether this definition is too all-encompassing to possess any practical utility. What useful common threads can be found in comparisons between the intricate architectures of epithelial cells and the bud processes of mating yeast? Although from the morphologic perspective such comparisons are difficult to support, they provide important insights into the common mechanistic elements which must be present in order for any cell to restrict a particular component of its surface membrane to a recognizable domain. While the precise biochemical identity of these elements may be variously determined by a cell's phylogenetic pedigree or differentiation program, their essence can be discerned in the production of any polarized phenotype.

For the purposes of this discussion it is useful to divide these elements into two categories—sorting signals and sorting machinery (Caplan and Matlin, 1989). We define a sorting signal as any information embedded within the primary, secondary, or tertiary structure of a given protein that somehow specifies the protein's appropriate localization. Sorting machinery comprises all of the cellular components and mechanisms that interpret a protein's sorting signal and act on that information in order to ensure that the protein accumulates at its site of ultimate functional residence. It must be emphasized that these definitions of sorting signals and sorting machinery make no presuppositions about their natures or modes of action. Whether a protein is vectorially targeted to a given domain during the course of its biosynthetic processing or is concentrated at a site solely by virtue of region-specific associations that increase its stability, a sorting signal embedded within that protein must have functioned to specify the necessary interactions. Similarly, whether the components that respond to a polypeptide's sorting signal include a receptor that mediates the protein's segregation as it passes through the Golgi complex or to the cytoskeletal elements that stabilize it in a plasmalemmal subdomain, sorting machinery must function to ensure the protein's nonuniform distribution. Although the precise composition of sorting signals and sorting machinery may vary widely among cell types, explorations of their structure and function can contribute to our understanding of fundamental issues in morphogenesis and cellular information processing.

Of all the polarized cell types alluded to in the preceding paragraphs, none has been (in the paraphrased words of e. e. cummings) "prodded by the naughty thumb of science" as thoroughly as epithelia. The appeal of epithelial cells for the study of polarity is the product of a number of factors. Foremost among these, of course, is their accessibility to experimental manipulation. Numerous cell culture lines have been developed which retain the dramatic polarity of their parent tissues when passaged *in vitro* (Caplan and Matlin, 1989; Simons and Fuller, 1985; Rodriguez-Boulan and Nelson, 1989). Furthermore, the ability of these cell lines to form occluding junctions and to be grown on permeable filter supports permits investigators to gain simultaneous and independent access to the two cell surface domains. Finally, the obvious relevance of polarity to these cells' physiologic functions, which include vectorial fluid and solute transport (Schultz, 1986), renders the mechanisms underlying their polarity all the more interesting.

A further inducement to study polarized epithelial cells was provided by the seminal observations of Rodriguez-Boulan and Sabatini (1978). These investigators studied Madin–Darby canine kidney (MDCK) cells that had been infected with enveloped RNA viruses. They found that the influenza virus assembled at and budded from the apical surface of these polarized cells, while the vesicular stomatitis virus (VSV) assembled at and budded from their basolateral membranes. Subsequent studies revealed that the spike glycoproteins that populate the viral envelopes accumulate at the plasmalemmal domain from which budding is to occur. Thus, influenza hemeagglutinin (HA) is targeted by infected cells to the apical membrane, while the VSV G-protein is sent to the basolateral surface (Rodriguez-Boulan and Pendergast, 1980). Transfection studies, employing MDCK cells that stably express these glycoproteins in the absence of any other viral components, provide similar results (Stephens *et al.,* 1986; Gottlieb *et al.,* 1986; Roth *et al.,* 1983; Jones *et al.,* 1985). Since it has been demonstrated that the viral proteins exploit the host cell's mechanisms for biosynthesis and postsynthetic processing, it was clear that the viral glycoproteins possess within their structure all of the information that is required to ensure their appropriate sorting and targeting. Researchers were thus provided with a model system which is accessible to molecular manipulations for the study of this sorting information.

In the decade since Rodriguez-Boulan and Sabatini's initial observations, a number of important studies have probed the sorting behavior
of proteins constructed of various portions and combinations of these
paradigmatic viral polypeptides in an effort to identify the biochemical
composition of the sorting information. A thorough treatment of the large
and complex literature which this work has spawned is beyond the scope
of this chapter. A number of review articles have undertaken a much
more thorough analysis of this field (Caplan and Matlin, 1989; Rodriguez-
Boulan and Nelson, 1989; Simons and Wandinger-Ness, 1990; Nelson,
1992). For the purposes of this discussion, it is important only to note
that while several general rules have emerged from the study of the viral
proteins, no sorting signals have been identified. While there are a variety
of possible sources for the difficulties and inconsistencies that have sur-
faced in this line of research, perhaps the most important relates to the
issue of tertiary structure (Caplan and Matlin, 1989).

Previous ruminations on the nature of sorting signals have pointed out
that these entities need not be composed of a single series of contiguous
amino acids but may instead be formed of several widely separated domains of a polypeptide which are brought into the appropriate apposition
by the acquisition of tertiary structure. Anything which perturbs protein
folding, therefore, may also impair the recognition of the embedded sorting
information. The production of deletion constructs, as well as the engineering of chimeras composed of portions of unrelated proteins, may result
in polypeptide products whose folding patterns bear little resemblance to
those of the parent molecules. Consequently, the sorting behavior of these
unnatural creations may not be interpretable by the cells in which they
are expressed. Since only the crudest estimates of tertiary structure can be
gathered for most membrane proteins, the degree to which these putative
peturbations interfere with the analysis of sorting signals cannot be determined readily.

II. SORTING SIGNALS: ION PUMPS IN POLARIZED EPITHELIAL CELLS

In order to avoid some of the potential difficulties enumerated above,
we have chosen to use transmembrane proteins belonging to the E_1–E_2
class of ion-transporting ATPases as substrates in our search for epithelial-
sorting signals. Perhaps the best-known delegate from this assembly of
plasmalemmal and organellar ion pumps is the Na,K-ATPase. The sodium
pump is a nearly ubiquitous component of the plasma membranes of
animal cells. By using energy in the form of ATP to defy steep concentration gradients, the Na,K-ATPase is able to move potassium ions into and

sodium ions out of cells, thus playing a critical role in maintaining membrane potential and cellular osmotic balance (Sweadner and Goldin, 1980). In most polarized epithelia the sodium pump is restricted in its distribution to the lateral and basal surfaces of the plasmalemma (Caplan, 1990). This configuration permits the ion gradients generated by Na,K-ATPase function to be exploited by a number of vectorial transport mechanisms.

The sodium pump's closest cousin in the E_1–E_2 family is the gastric H,K-ATPase. This pump carries out the ATP-driven exchange of extracellular potassium for intracellular protons in order to acidify the lumen of the stomach (De Pont et al., 1988). In its native parietal cells the H,K-ATPase inhabits the apical membrane as a well as a population of intracellular vesicles, termed tubulovesicular elements, which fuse with the apical surface on the secretagogue-mediated stimulation of acid secretion (Smolka and Weinstein, 1986). Structurally and functionally the H,K-ATPase and the Na,K-ATPase are highly homologous. Both pumps are heterodimers composed of a 100-kDa α-subunit and a heavily glycosylated 55-kDa β-subunit (Jorgensen, 1982; Okamoto et al., 1990). At the amino acid level the H,K- and Na,K-ATPase α-subunits are ~65% identical, while the β-subunits share ~40% identity (see Fig. 1) (Reuben et al., 1990; Shull, 1990; Shull and Lingrel, 1986). The mechanisms of the two pumps are remarkably similar, involving in both cases a complex series of conformational transitions, the formation of a phosphorylated intermediate, and sensitivity to vanadate inhibition (De Pont et al., 1988).

Given the dramatic homologies linking these two pump complexes, it is fascinating from a cell biologic perspective that they are sorted to opposite domains of polarized epithelial cells. This divergent sorting behavior is elegantly manifest in gastric parietal cells. While the H,K-ATPase is concentrated in apical and preapical compartments, the sodium pump exhibits its characteristic basolateral localization (see Fig. 2) (Soroka et al., 1993). Despite their similarity, therefore, these two ATPases can be distinguished by the sorting machinery operating in the parietal cell, which is thus able to mediate their segregation to separate domains. We have undertaken to exploit this phenomenon in order to try to identify those portions of the pump molecules which confer their distinct sorting properties. By using closely related proteins in the generation of chimeric constructs, we hope to avoid the potential for serious disruptions in tertiary structure discussed above. Furthermore, by assaying the catalytic activity of the resultant constructs, it should be possible to demonstrate that their structures are sufficiently intact to carry out the intricate sequence of conformational changes required by the enzymatic cycles of the parent pumps.

The relatively recent cloning of cDNAs encoding all of the Na,K-ATPase and H,K-ATPase α (Shull and Lingrel, 1986) and β-subunits

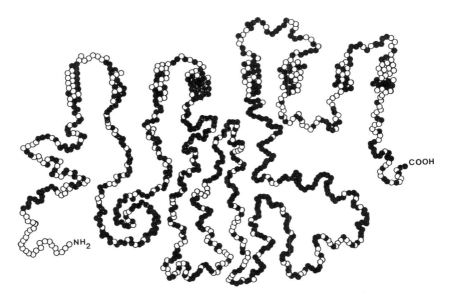

FIGURE 1   The α-subunits of the H,K- and Na,K-ATPase are ~65% identical. Each circle in the diagram represents an amino acid in the H,K-ATPase α-subunit sequence. Filled circles correspond to residues shared between the Na,K- and the H,K-ATPase α-subunit sequences. Helices represent putative transmembrane domains. The small loops face the extracellular space, while the large loops are cytoplasmic. (Reproduced from the *Journal of Cell Biology*, 1993, vol. 121 pp. 283–293, by copyright permission of the Rockefeller University Press.)

(Reuben *et al.*, 1990; Shull, 1990) has made it possible to express these proteins by transfection in epithelial cell lines. It should thus be fairly straightforward to analyze the contribution of each subunit to the biogenesis and sorting of the two pump systems. We are grateful to a number of investigators who have generously made these cDNAs available to us: Gary Shull (H,K-ATPase α-subunit), Michael Reuben and George Sachs (H,K-ATPase β-subunit), and Edward Benz, Jr. (Na,K-ATPase α- and β-subunits). For all of the transfection work described in this chapter cDNAs encoding pump subunits or chimeric constructs were subcloned into the pCB6 vector (kind gift of Michael Roth) (Brewer and Roth, 1991). This vector uses a CMV promoter to drive insert expression in mammalian cells. All chimeras were prepared and analyzed as inserts in the BlueScript plasmid prior to insertion into pCB6.

Before initating expression studies it was necessary to generate antibodies that can specifically detect each pump molecule. Given their high degree of sequence identity, some care was required in order to create

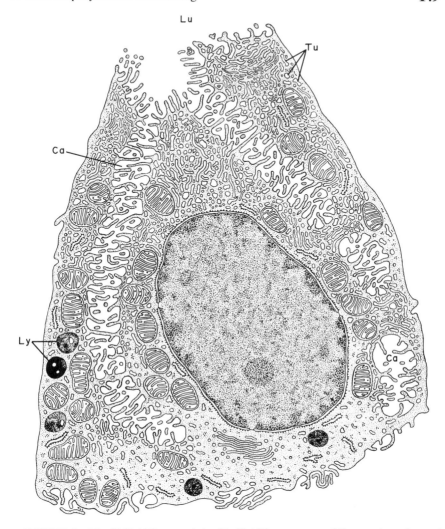

FIGURE 2 The H,K-ATPase and the Na,K-ATPase occupy different plasmalemmal domains in gastric parietal cells. The fine structure of the gastric parietal cell is presented. Lu, lumen of the gastric gland; Tu, tubulovesicular elements; Ca, secretory canaliculi; Ly, lysosomes. The H,K-ATPase is concentrated in the tubulovesicular elements, which serve as an intracellular storage compartment for H,K-ATPase and the secretory protein intrinsic factor. On stimulation, the tubulovesicular elements fuse with the apical surface to form the secretory canaliculi. The Na,K-ATPase is limited in its distribution to the basolateral surfaces. (Figure adapted with permission from *Cell Fine Structure,* by T. Lentz, 1971, W. B. Saunders.)

probes that do not cross-react with antigenic determinants common to both the H,K- and the Na,K-ATPases. The importance of this consideration is underscored by the fact that the polarized cell lines available for transfection and expression studies possess a large complement of endogenous Na,K-ATPase.

For the studies described in this chapter we have prepared two polyclonal antibodies directed against synthetic peptides whose sequences are derived from portions of the H,K-ATPase α-subunit that lack homology to the corresponding domains of the Na,K-ATPase α protein. Both of these antibodies react with rat gastric H,K-ATPase by Western blot analysis. Furthermore, in immunofluorescence experiments both probes react strongly with parietal cells to produce a pattern characteristic of the H,K-ATPase when they are applied to sections of fixed rat stomach. Neither antibody exhibits any affinity for the Na,K-ATPase as judged by Western blot, immunoprecipitation, and immunocytochemistry (Gottardi and Caplan, 1993a; Okusa *et al.*, 1994).

We chose to analyze sorting of the ion pumps in LLC-PK1 cells (kind gift of Kurt Amsler). This cell line is derived from the pig kidney proximal tubule and, like its parent tissue, is graced with a lavish apical brush border (Pfaller *et al.*, 1990). These cells are polarized in culture and are capable of transepithelial fluid transport as revealed by their propensity to form domes. Their utility for our studies derives from their relatively easy transfectability, their capacity for high levels of exogenous protein production, and their columnar morphology, which facilitates determination of the surface distribution of the introduced molecules.

Our initial experiments examined the fate of H,K-ATPase α-subunit expressed in the absence of its appropriate β component. When the H,K-ATPase α-subunit is expressed in LLC-PK1 cells, Western blot analysis demonstrates that the cells produce a polypeptide of the appropriate molecular mass ($\sim$100 kDA) that is reactive with our H,K-ATPase antibodies. Examination of the transfected cells by immunofluorescence using these antibodies reveals a pattern characteristic of the endoplasmic reticulum (ER). Similar results are obtained when the H,K-ATPase α-subunit is expressed by transient transfection in COS cells (Gottardi and Caplan, 1993b). This localization is not entirely surprising, since a large body of data has been gathered demonstrating that multimeric proteins that are prevented from attaining their appropriate quaternary structure are specifically retained in the ER, where they await completion of their assembly or targeting to a degradative pathway (Rose and Doms, 1988). Furthermore, by individually expressing Na,K-ATPase subunit proteins, several investigators have shown that the α-subunit requires interaction with the β polypeptide in order to depart the ER and reach the cell surface (McDonough *et al.*, 1990). Our observations suggest that, by itself, the H,K-

ATPase α-subunit is unable to fold or assemble correctly and is thus retained in the membranes of the ER.

The situation is quite different when the H,K-ATPase α-subunit is expressed in the company of the H,K-ATPase β-subunit. When immunolocalization is performed on LLC-PK1 cells coexpressing both H,K-ATPase subunit polypeptides, the α-subunit is seen to densely populate the apical surface. No basolateral and very little intracellular α staining are detected. Double labeling with antibodies directed against the Na,K-ATPase α-subunit reveals that the sodium pump retains its exclusively basolateral localization. It appears, therefore, that the LLC-PK1 cells are able to recapitulate the gastric parietal cell's sorting accomplishments; they are able to distinguish between the H,K-ATPase and the Na,K-ATPase and to concentrate them at different plasmalemmal domains (Gottardi and Caplan, 1993a).

In the context of these results, it is interesting to remember that the LLC-PK1 cells have a large and actively turning over complement of Na,K-ATPase. They are constitutively synthesizing sodium pump α- and β-subunits at rates that are comparable to or larger than the rate at which transfected cells translate H,K-ATPase α-subunit. It would appear, therefore, that although an ample pool of newly synthesized sodium pump β-subunit is available in the rough ER of LLC-PK1 cells, H,K-ATPase α-subunit does not assemble with it. The presence of H,K-ATPase β-subunit is necessary and sufficient to permit the H,K α to assemble correctly and depart the ER. This surprising degree of selectivity suggests that the domains of the Na,K-ATPase and H,K-ATPase α-subunits that interact with their respective β's are quite distinct from one another (Gottardi and Caplan, 1993b). As seen below, our chimera studies have confirmed this conjecture and provided some information as to where in the primary structures of the α-subunits those interaction domains may reside.

Close examination of the cells expressing both H,K-ATPase subunits led to a rather surprising observation. While the H,K α-subunit was restricted in its distribution to the apical surface, the corresponding β-subunit was present both at the apical surface and in a population of subapical vesicles. The Na,K α could not be detected in this compartment. As would be expected of an intracellular population, the H,K β residing in these structures was inaccessible to antibody added to the apical surface of cells which had been cooled to 4°C prior to fixation. These findings suggest that the H,K β can exist outside of the endoplasmic reticulum in the absence of either α-subunit. In order to further characterize this behavior as well as the nature of this vesicular compartment, we generated LLC-PK1 cells that express H,K β in the absence of H,K α (Gottardi and Caplan, 1993a).

Analysis of the cell line transfected with the H,K β alone reveals that this protein accumulates at the apical surface as well as in the intracellular compartment discussed above (see Fig. 3). The sodium pump α-subunit retains its normal basolateral distribution. Immunoprecipitation from transfected cells labeled with [^{35}S]methionine fails to demonstrate any endogenous LLC-PK1 polypeptides that specifically associate with the H,K β protein under these conditions. These data suggest that the H,K β's departure from the ER is not dependent on assembly with an α-subunit. They further indicate that the H,K β encodes information sufficient to ensure its apical targeting. It is tempting to conclude, therefore, that sorting signals encoded by the H,K β are necessary and sufficient to ensure the apical localization of the H,K-ATPase holoenzyme.

In order to test the hypothesis that the H,K β is responsible for the H,K pump's apical sorting, we have generated a chimeric α-subunit which comprises the NH$_2$ terminal 519 amino acids of the H,K α and the COOH terminal 500 amino acids of the Na,K α. The resulting polypeptide, re-

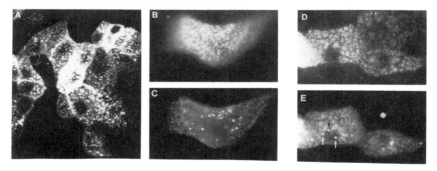

FIGURE 3   The H,K-ATPase β-subunit accumulates at the apical plasmalemma and in subapical endosomes in transfected LLC-PK1 cells. LLC-PK1 cells singly transfected with a cDNA encoding the H,K-ATPAse β-subunit (A,B,C) or doubly transfected with cDNAs encoding the H,K-ATPase α- and β-subunits (D,E) were examined by indirect immunofluorescence using antibodies directed against the H,K-ATPase β-subunit (A,B,C,E) or the H,K-ATPase α-subunit (D). When the cells are examined *en face*, the H,K-ATPase β-subunit displays a fine punctate pattern characteristic of apical labeling in both singly and doubly transfected cells. When the plane of focus is moved from the surface to the interior of the cells, intracellular vesicular staining becomes apparent. The presence of vesicular labeling is independent of H,K-ATPase α-subunit expression. Furthermore, the α-subunit is present only at the cell surface and does not accompany the β-subunit into the vesicles (compare D to E). Reproduced from the *Journal of Cell Biology*, 1993, vol. 121, pp. 283–293, by copyright permission of the Rockefeller University Press.)

ferred to as H519N, represents the first four predicted transmembrane domains of H,K α fused to the final four predicted transmembrane domains of Na,K α. The fusion point occurs in the putative ATP-binding site, which is identical in the two pumps, and regenerates this sequence exactly.

When expressed by itself in LLC-PK1 cells the H519N chimera is able to depart the ER and accumulates at the apical plasmalemma. Immunolocalization of the Na,K-ATPase subunits reveals that, while the α is restricted to the basolateral surface, the β-subunit is distributed over both domains of the plasma membrane (see Fig. 4) (Gottardi and Caplan, 1993a). It would appear, therefore, that the H519N chimera can assemble with the Na,K β-subunit and that this complex can depart the ER. Since the full-length H,K α is not similarly capable of interacting with Na,K β, we conclude that the COOH terminal half of the α-subunit specifies its preference in β-subunits. This interpretation has been confirmed through transient transfection experiments involving the complementary α chimera, N519H, which assembles exclusively with the H,K β (Gottardi and Caplan, 1993b).

The fact that the heterodimer composed of this chimera and the normally basolateral Na,K β-subunit behaves as an apical membrane protein strongly indicates that sorting information is not the exclusive province of the β-subunits. These results demonstrate that the NH_2 terminal half of the α-subunit must be endowed with a dominant sorting signal. Our data are equally compatible with either or both of the following two possibilities: (1) the NH_2 terminal portion of the H,K α directs apical sorting or (2) the corresponding portion of the Na,K α functions to mediate targeting to the basolateral surface. Experiments employing the complementary chimera, N519H, will allow us to distinguish between these alternatives. These chimera results, taken together with the H,K β studies, demonstrate that both subunits contribute to the sorting of the H,K-ATPase. The signifigance of this apparent redundancy is discussed below.

In order to understand the role of the H,K-ATPase β-subunit in sorting it was necessary to characterize the intracellular compartment in which it accumulates. Examination at the electron microscopic level of cells expressing solely H,K β reveals that this protein is present at the apical plasmalemma, in elements of the rough ER, and in structures reminiscent of tubular endosomes. The presence of H,K β in endosomes was confirmed by a double-labeling experiment employing horseradish peroxidase (HRP) as a marker for fluid-phase endocytosis. As early as 5 min after the addition of HRP this tracer could be detected in intracellular compartments which also stained positively for the H,K β. We conclude, therefore, that this protein is capable of entering the LLC-PK1 cell's endocytic pathway (Gottardi and Caplan, 1993a).

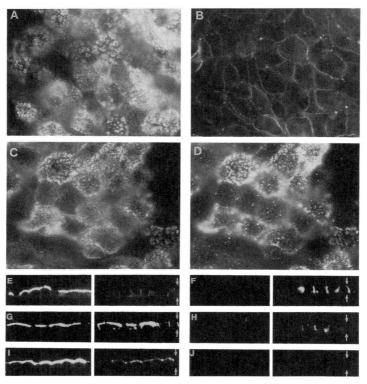

FIGURE 4 The H519N chimera acculates at the apical surface of LLC-PK1 cells in association with endogenous Na,K-ATPase β-subunit. LLC-PK1 cells transfected with the cDNA encoding H519N (A–E, G, I) and untransfected cells (F,H,J) were examined by indirect immunofluorescence (A–D) or by confocal microscopic XZ sectioning (E–J). When examined *en face* with the antibody directed against the H,K-ATPase α-subunit a typical apical staining pattern is observed (A,C). The Na,K-ATPase α-subunit appears to be localized to the basolateral surface (B), while the Na,K-ATPase β-subunit distribution appears to be primarily apical (D). Analysis of cross sections generated by confocal microscopy reveals that H519N is, in fact, restricted to the apical surfaces of transfected cells (E,G,I, left) and absent from untransfected cells (F,H,J, left). While the Na,K-ATPase α-subunit is basolateral in both transfected and untransfected cells (compare the right panels of E and F), the Na,K-ATPase β-subunit is apical and lateral in transfected cells but only basolateral in untransfected controls (compare the right panels of G and H). The presence of the Na,K-ATPase β-subunit in the apical membrane is confirmed by adding Na,K-ATPase antibody to the media bathing the apical surfaces of unpermeabilized monolayers. The apical membrane is labeled in transfected (I, right) but not untransfected cells (J, right). (Reproduced from the *Journal of Cell Biology*, 1993, vol. 121, pp. 283–293, by copyright permission of the Rockefeller University Press.)

The presence of H,K β in endosomes raises a number of interesting questions. Why, for example, is only β found in the endosomes of cells expressing both H,K-ATPase subunits? What structural determinants of the H,K β polypeptide allow this protein to participate in endocytic internalization? Is this capacity for endocytosis relevant to the normal function of the H,K β in gastric parietal cells and is it involved in polarized sorting? It should surprise no one that definitive answers to these questions have yet to be elucidated. It should also come as no surprise that this lack of definitive information does not prevent us from availing ourselves of this opportunity to speculate.

The absence of H,K α from the endosomal compartment suggests that while the β-subunit can participate in endocytosis, the holoenzyme does not share this property. Two models can be proposed to explain this behavior. Assembly with the α-subunit must either mask the β-subunit's affinity for endocytosis or actively impede the β's incorporation into coated pits and endosomes. We favor the second of these alternatives. It has been established that the Na,K-ATPase α-subunit participates in specific interactions with elements of the submembranous cytoskeleton (Nelson and Veshnock, 1987; Nelson and Hammerton, 1989; see also Nelson *et al.*, this volume). The association between the sodium pump α-subunit and ankyrin is thought to link the holoenzyme to the cytoskeletal meshwork and thus substantially retard its internalization and degradation (Hammerton *et al.*, 1991). Recently, similar associations have been demonstrated for the H,K-ATPase. The H,K α-subunit appears to interact with ankyrin and may thus be subject to the stabilizing influences of the cytoskeletal scaffolding (Smith *et al.*, 1993). If the H,K-ATPase α-subunit is tethered to the LLC-PK1 cell's endogenous ankyrin–fodrin network, it may be prevented from interacting with the endocytic machinery. H,K β polypeptides that participate in stable heterodimeric complexes may thus be barred from internalization. Those unassembled β-subunits expressed at the surface would not be subject to this restraint and would be available to the cell's endocytic apparatus.

Close examination of the H,K ATPase β-subunit's amino acid sequence reveals the presence of a four amino acid motif which is likely to account for this protein's capacity for endocytosis. It has been demonstrated that a large number of proteins which are subject to rapid internalization share a structural feature consisting of a tight β turn which includes a tyrosine residue (Goldstein *et al.*, 1985). Site-directed mutagenesis experiments further demonstrate that this element comprises a coated pit localization signal. Proteins endowed with this signal are rapidly internalized via coated pits and delivered to early endosomes. Disruption of this signal produces a protein which does not participate in endocytosis. The transferrin receptor exemplifies this behavior. The cytoplasmic tail of this protein contains the sequence YTRF which has been shown to form a tight β turn as well as to specify the receptor's incorporation into coated pits (Collawn *et al.*, 1990). Site-directed mutagenesis reveals that any sequence of the form YXRF, presented forward or backward, will also satisfy both of these criteria (Girones *et al.*, 1991). The cytoplasmic tail of the rabbit H,K-ATPase β-subunit contains the sequence FRHY. This motif is present as well in the sequences of H,K β's cloned from human, rat, and pig (Reuben *et al.*, 1990; Shull, 1990). None of the Na,K-ATPase β-subunit isoforms are similarly endowed. It would appear, therefore, that the H,K-ATPase β-subunit is included in the endocytic pathway by virtue of a putative coated pit localization signal. Site-directed mutagenesis experiments designed to test this hypothesis are currently underway.

As mentioned above, the H,K-ATPase is not a permanent resident of the apical plasma membrane of its native gastric parietal cell. The enzyme is concentrated in a large population of intracellular tubulovesicular elements that fuse with the apical surface on stimulation. The H,K-ATPase is retrieved from the plasmalemma and returned to its intracellular compartment on cessation of the stimulus. Little is currently known of the mechanisms through which these insertion and recovery processes are governed. It is interesting to speculate that the process of reinternalization may be achieved through coated pit-mediated endocytosis. According to this model, the YXRF sequence embedded in the β's cytosolic tail would function to direct the H,K-ATPase into the endocytic pathway and ensure its removal from the cell surface. In order to be applicable to the gastric parietal cell, however, this model must also explain why this endocytosis signal results in H,K-ATPase internalization only after the impetus for acid secretion has subsided. It has recently been shown that the interaction between the H,K-ATPase α-subunit and the gastric parietal cell's cytoskeleton is dynamically regulated. In unstimulated stomach, ankyrin and fodrin are distributed throughout the cytosol, whereas in active acid-secreting cells these proteins are closely adherent to the membranes of the secretory canaliculi (Mercier *et al.*, 1989). It is possible that the H,K-ATPase of stimulated parietal cells is firmly attached to a protein network which precludes its participation in the endocytic pathway. Removal of the stimulus results in the dissolution of this network, thus allowing the β-subunit's signal to exert its influence. This scheme is, of course, entirely hypothetical. Its verification awaits the development of a cell culture or transgenic animal model in which the biology of the gastric parietal cell can be subjected to molecular analysis.

Only the possible role the H,K β's proclivity for endocytosis may play in this protein's apical localization remains to be discussed. Studies of protein sorting in MDCK cells have revealed that endocytosis signals are also sufficient to specify basolateral targeting. This contention receives elegant support from site-directed mutagenesis experiments performed on the influenza HA protein. This normally nonendocytosed apical protein can be induced to enter the endocytic pathway by replacing a single amino acid in its cytosolic tail with a tyrosine residue (Brewer and Roth, 1991). This modification creates a tyrosine-containing β turn motif and produces a protein characterized by rapid internalization. When expressed in MDCK cells, this variant HA protein is delivered to and accumulates at the basolateral surface. Similar results have now been generated for a number of proteins.

In the context of this relationship between endocytosis signals and basolateral sorting, it is surprising that the H,K-ATPase β-subunit is a component of the apical membrane of transfected LLC-PK1 cells. We believe that this behavior reflects an interesting difference between LLC-PK1 and MDCK cells that may shed light on the involvement of endocytosis signals in polarized sorting. While MDCK cells are derived from the canine renal distal nephron, the LLC-PK1 line traces its origins to the proximal tubule of the pig kidney. The mammalian proximal tubule is specialized to carry out a number of transport functions which serve to reclaim the vast majority of solutes and fluid present in the glomerular filtrate (Mandel and Balaban, 1981). Peptides and small proteins are recovered from the tubule fluid via coated pit-mediated endocytosis. The proximal tubule's adaptation to this task is demonstrated through morphologic examination, which reveals that the apical membranes of the epithelial cells are richly endowed with coated pits and endocytic profiles (Rodman et al., 1984). The basolateral surfaces of these cells are much less extravagently adorned with endocytic machinery. The situation is quite different in distal segments of the nephron, where basolateral endocytosis appears to predominate.

Our results are consistent with the possibility that endocytosis signals specify sorting to functional rather than to topographic domains. According to this model, coated pit localization sequence-mediated sorting ensures a protein's delivery to the cell's most endocytically active plasmalemmal surface. In the case of MDCK cells, this would appear to correspond to the basolateral membrane. In contrast, LLC-PK1 cells appear to retain the proximal tubule's propensity for apical endocytosis. According to this scheme, the apical localization of the endocytosis signal-bearing H,K β in LLC-PK1 cells reflects the apical predominance of these cells' endocytic machinery. This model predicts that the H,K β should be concentrated at the basolateral surfaces of MDCK cells, while the tyrosine mutant of the influenza HA protein should behave as an apical protein in LLC-PK1 cells. Preliminary data from experiments designed to test these predictions suggest that, in fact, both are true. It remains to be determined how the cell establishes its endocytically active domains and how the endocytic and sorting machinery interact to concentrate internalizable proteins at these sites.

III. EPITHELIAL AND NEURONAL SORTING: SIMILARITIES AND DIFFERENCES

Recently, several groups have begun to extend the insights gained through the study of sorting in epithelia to nonepithelial cell types. Studies on cultured hippocampal neurons have led to the suggestion that neuronal sorting mechanisms may be closely related to those employed by epithelial cells. Infection of these cells with enveloped viruses revealed that influenza HA protein accumulates in axonal processes, while the VSV G-protein is restricted to the dendrites (Dotti and Simons, 1990). Further experiments indicated that glycosylphosphatidylinositol (GPI)-linked proteins, which are restricted to the apical surfaces of polarized epithelial cells (Lisanti *et al.*, 1990), are predominantly axonal in neurons (Dotti *et al.*, 1991). These observations have prompted the speculation that the same signals and machinery involved in sorting to the basolateral and apical domains of epithelia function in neurons to mediate targeting to dendrites and axons, respectively. In order to examine this hypothesis further, we tested the prediction that axonal proteins would be apically sorted when expressed by transfection in epithelial cells. Our initial studies have made use of the Na,Cl-dependent GABA transport system which populates the presynaptic plasma membranes of GABA-ergic nerve terminals.

GABA-ergic neurons make use of transmitter reuptake in order to terminate synaptic signaling (Iversen, 1971). This uphill reabsorption is accomplished by a Na,Cl-dependent cotransport system (Radian and Kanner, 1983). Similar cotransport activities, with related pharmacologic and physiologic properties, function at serotonergic, noradrenergic, and dopaminergic terminals. Recently, cDNA clones encoding each of these reuptake systems have been isolated (Guastella *et al.*, 1990). Sequence analysis reveals that they are closely interrelated, with an average similarity at the amino acid level of >60% and superimposable hydropathy plots. Cloning has further revealed that the transport system for betaine, which is involved in osmoregulation in the kidney, is also a member of this family and shares the same degree of homology (Yamauchi *et al.*, 1992). The remarkable structural similarity between the GABA and betaine transport systems is underscored by the fact that GABA is a high-affinity substrate for the betaine carrier (Yamauchi *et al.*, 1992).

We have transfected polarized epithelial cells with a cDNA encoding the GABA transporter. Immunocytochemical and transport studies reveal that this protein is sorted to the apical plasmalemma. This observation suggests that models relating apical and axonal sorting mechanisms may indeed be valid. More importantly, however, it creates the opportunity to pursue new studies of sorting signals and transport function. Transport measurements reveal that the osmotically inducible betaine system is localized to the basolateral surfaces of renal epithelial cells (Yamauchi *et al.*, 1991). Thus, the sorting information embedded in the structure of the GABA transporter must differ from that associated with its close cousin. We are in the process of generating GABA/betaine transporter chimeras in order to determine where within the structures of these proteins their sorting signals reside.

We wondered whether the equation between neuronal and epithelial sorting mechanisms would hold in the case of the Na,K-ATPase. Since this protein is basolateral in epithelia, the model predicts that the sodium pump should be restricted to dendritic membranes. On the basis of physiologic considerations, however, we expected that Na,K-ATPase must be present in the membranes of both axons and dendrites. We wished to determine, therefore, whether the sodium pump needs of these two compartments might be met by two distinct isoforms of the sodium pump α-subunit.

Three isoforms of the α-subunit of the Na,K-ATPase have been identified and cloned (Shull *et al.*, 1986). The α1 protein appears to be present in most cell types; while α2 is restricted to muscle, fat, and glia; and α3 is expressed predominantly in neurons (McGrail *et al.*, 1991). We generated isoform-specific synthetic peptide antibodies and used them to localize the Na,K-ATPase in hippocampal neurons in culture as well as *in situ* (Pietrini *et al.*, 1992). We find that two isoforms of Na,K-ATPase, α1 and α3, are simultaneously expressed in hippocampal neurons in culture. In order to assess the polarity of the cultured neurons we used a polyclonal antibody directed against the microtubule-associated protein 2 (MAP2), a cytoskeletal protein restricted to the somata and dendrites of hippocampal neurons (Caceres *et al.*, 1986), as a dendritic marker. A monoclonal antibody directed against growth-associated protein 43 (GAP43), which is preferentially distributed to the axonal domain (Goslin *et al.*, 1990), was used to mark axons. Double-staining experiments employing these probes reveal a clearly polarized, complementary distribution of the two proteins at all the ages of culture analyzed.

When neurons were labeled with the monoclonal antibody directed against α1, virtually every neuronal process visible by phase-contrast microscopy was stained (see Fig. 5). In double-label experiments, α1 immunostaining is associated with both MAP2-positive and MAP2-

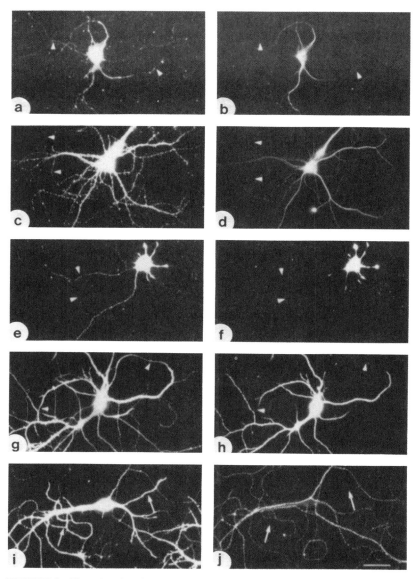

FIGURE 5 The α1 and α3 isoforms of the Na,K-ATPase are present in both the axons and dendrites of hippocampal neurons in culture. Hippocampal neurons grown in culture for 7 (a,b,e,f), 14 (g–j), or 18 days (c,d) were examined by double-label immunofluorescence using antibodies directed against α1 (a,e) or α3 (c,g,i) in conjunction with probes directed against the dendritic marker MAP2 (b,d,f,g) or the axonal marker GAP43 (j). As described in the text, both α-isoforms are present in both MAP2-positive dendrites and MAP2-negative axons. The scale bar = 16 μm. (From Pietrini *et al.*, 1992).

negative processes. The labeling of MAP2-negative processes clearly demonstrates the presence of α1 immunoreactivity in axons. Immunostaining with the polyclonal antibody selective for the α3 isoform demonstrates that it also has an unrestricted distribution in hippocampal neurons. α3 immunoreactivity was detected in both MAP2-positive and MAP2-negative processes. The latter result demonstrates that the α3-isoform is present in axons. To rule out the possibility that this isoform is exclusively associated with axons, cultures were double-stained with antibodies against α3 and GAP43 (Fig. 5 i,j). All processes stained with GAP43 were also positive for α3, but some α3-positive processes were GAP43-negative. The latter processes had the distinctive morphology of dendrites. These results clearly demonstrate that the α3-isoform, like α1, is present in both dendrites and axons.

The observations presented above demonstrate that neurons express α3 Na,K-ATPase in both dendrites and axons. The sorting behavior of this isoform in polarized epithelia, however, has not previously been examined. It was necessary to determine, therefore, whether the unpolarized expression of α3 reflects its behavior in epithelia. Toward this end, we have stably transfected LLC-PK1 cells with the cDNA encoding α3 (see Fig. 6). Immunofluorescence analysis of these cells reveals a basolateral staining pattern for α3. The exclusively basolateral localization has

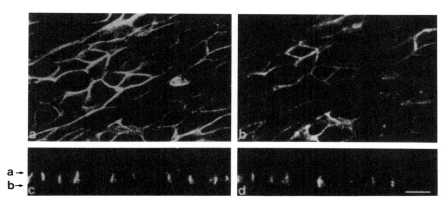

FIGURE 6 The α1- and α3-isoforms of the Na,K-ATPase are basolateral in LLC-PK1 cells. LLC-PK1 cells transfected with the cDNA encoding the α3-subunit of Na,K-ATPase were examined by double-label immunofluorescence confocal microscopy using antibodies directed against α1 (a,c) and α3 (b,d). When viewed *en face* (a,b) or in XZ cross sections (c,d) both the exogenous α3 and the endogenous α1 proteins were distributed at the basolateral surfaces. Apical and basolateral surfaces in the XZ cross sections are indicated by the arrows labeled a and b, respectively. The scale bar = 15 μm. (From Pietrini *et al.*, 1992).

been confirmed by confocal imaging of cells transfected with α3 and double-labeled with α1 antibody.

Based on the observations presented above, it is tempting to conclude that the mechanisms which mediate sodium pump accumulation in epithelia are not identical to those which function in polarized neurons. Recent evidence suggests, however, that this interpretation might be overly restrictive. As has been mentioned above, the Na,K-ATPase appears to interact with elements of the cytoskeleton which underlie the basolateral surfaces of polarized MDCK cells. Hammerton *et al.*'s studies on the initial targeting of the Na,K-ATPase suggest that these interactions may play an important role in determining the Na,K-ATPase's polarized distribution (Hammerton *et al.*, 1991). These investigators used a cell surface labeling technique to follow the cell surface delivery of newly synthesized sodium pump in MDCK cells. They found that this protein was delivered to the apical and basolateral surfaces in roughly equal quantities. They also found that the stabilities of the apical and basolateral sodium pump populations were markedly different. While the $t_{1/2}$ of the basolateral Na,K-ATPase exceeded 24 hr that of the apical contingent was less than 90 min. These authors conclude that the polarized distribution of the sodium pump is not the product of sorting in the biosynthetic pathway. Instead, they attribute this protein's localization to the differential availability of stabilizing cytoskeletal interactions at the two epithelial surface domains.

In contrast to polarized epithelial cells, elements of the cytoskeleton appear to line the entire neuronal plasmalemmal surface. Isoforms of ankyrin and fodrin can be found throughout both axons and dendrites (Kordely and Bennett, 1991). Thus, if neuronal Na,K-ATPase were distributed according to the stabilization model discussed above it might be expected to be present in the membranes of both classes of processes. It is possible, therefore, that the sorting mechanisms that function in neurons and epithelia are extremely similar to one another. The equation relating the ultimate distributions produced by these mechanisms, however, may depend on which of a multiplicity of possible sorting mechanisms is relevant to a particular protein's targeting. A clearer understanding of the common features shared by neuronal and epithelial sorting must await a molecular analysis of the cellular components which mediate these processes.

IV. SORTING MACHINERY: A GENETIC APPROACH

Until recently, most polarity research has focused exclusively on investigations into the nature of sorting pathways and sorting signals. Few methods have been developed that can reliably and convincingly identify a priori the cellular proteins which are necessary for the generation and maintenance of the polarized state. The identification of components of the sorting machinery has been hampered by the lack of a suitable *in vitro* system capable of performing sorting functions. Without such an experimental paradigm, it is difficult to establish that proteins isolated via standard biochemical methods are important for or even relevant to the process of sorting. In an effort to overcome this limitation, we have begun to explore the possibility of applying genetic methods to this problem.

Our strategy makes use of the unique opportunities for combining genetic and molecular techniques afforded by P element insertional mutagenesis in *Drosophila*. Within 3 hr of emerging from its mother, the embryonic fruit fly is composed primarily of well-developed and easily recognizable epithelial structures. Immunocytochemical studies carried out in several laboratories have established that these epithelial tissues manifest dramatic biochemical polarity. Using antibodies directed against membrane proteins such as the Na,K-ATPase, fasciclin III, and Toll, it can be shown that these proteins enjoy polarized distributions in the *Drosophila* embryonic epithelia throughout ontogeny. It would appear, therefore, that the morphologic and biochemical correlates of *Drosophila* epithelial polarity appear to closely resemble those of vertebrate epithelia.

We are developing a screening protocol which will allow us to search for mutations that perturb the normal polar organization of embryonic epithelia. The approach we have chosen is predicated on the findings of Brown *et al.* (1989). These investigators have studied the sorting of placental alkaline phosphatase (PLAP) expressed by transfection in MDCK cells. PLAP is a GPI-linked protein and accumulates at the apical surfaces of polarized epithelia. A chimeric PLAP construct, composed of the PLAP ectodomain linked to the VSV G-protein's transmembrane and cytosolic domains (PLAPG), is sorted to the MDCK basolateral surface. Both versions of PLAP maintain alkaline phosphatase activity. We hypothesize that PLAP and PLAPG will manifest similar sorting behavior when expressed in the epithelia of the *Drosophila* embryo.

In the intact dechorionated embryo only the apical surface of the cellular blastoderm is exposed to the bathing medium. It should be possible, therefore, to histochemically stain for the apical presence of alkaline phosphatase activity in order to assess the epithelial polarity of embryos expressing either of these constructs. This would be a rapid and easy screen which could be applied to large numbers of flies. Mutant stocks to be screened by this method will express the relevant constructs as a result of genetic crosses with transgenic stocks generated by germ line transformation. Further characterization of candidate epithelial mutant stocks will be accomplished by immunostaining of embryos with antibodies directed against the Na,K-ATPase, fasciclin III, and Toll. By comparing the local-izations of these antigens in potential mutant embryos with their distribu-tions in wild-type embryonic epithelia it should be possible to confirm the identification of *Drosophila* mutants manifesting abnormalities of epithelial polarity.

Our ability to carry out our studies on early embryos will greatly facili-tate our search for polarity mutants, since any disruption of the genes required to establish polarized epithelia is likely to produce a phenotype which is lethal during the embryonic or larval stages. We are aided in our efforts by the relatively small number of unique genes associated with lethal mutations in the *Drosophila* genome (~5,000–10,000), which will make it possible to employ both the histochemical and immunocytochemi-cal techniques described. Finally, and most importantly, our work will rely on mutant stocks produced through the insertion of P elements bearing enhancer traps (Cooley *et al.*, 1988). This technology will permit us to limit our search to stocks that carry lethal mutations of genes which have been shown, by analysis of enhancer trap expression, to be expressed in polarized epithelial cells. We will begin our screen with these stocks, which can be thought of as "likely candidates" for harboring mutations relevant to epithelial polarity. This will greatly reduce the number of stocks we will need to screen and consequently will enhance our ability to identify interesting phenotypes. Once we identify lethal mutations asso-ciated with missorting, we will be able to use the special features of P elements to rapidly isolate and clone the responsible genes.

We have succeeded in generating the transgenic fly stocks necessary to carry this screen forward. Biochemical experiments employing phos-pholipase C (PLC) reveal that PLAP is anchored to the membrane via a GPI linkage when expressed in *Drosophila*. As would be expected, PLAPG remains resistant to PLC cleavage. Preliminary immunolocaliza-tion of PLAP and PLAPG in embryonic fly tissues suggests that their sorting in these cell types mirrors that observed in MDCK cells. PLAPG is concentrated in the basolateral membranes of embryonic epithelia and larval imaginal disks. PLAP appears to be apically distributed in these same structures. We are currently working to determine whether the histochemical method outlined above will be practicable. If so, we hope to be able to begin the process of identifying and characterizing the cellular elements required to generate and maintain the polarized state.

V. CONCLUSIONS

The preceding discussion represents an extremely idiosyncratic view of some of the questions which currently occupy researchers in the field of cellular polarity. If nothing else, it should be clear that progress toward answering these questions is still quite embryonic. We know almost nothing about the signals, mechanisms, and machinery which cells use to generate anisotropic distributions of plasmalemmal proteins. It is hopefully equally clear, however, that progress in this field will be relevant to a number of important problems in cell biology. A more complete understanding of membrane protein sorting will shed light not only on the regulation of epithelial membrane transport, but also on fundamental issues in cellular information processing.

Acknowledgments

We thank the Caplan lab group for helpful suggestions and comments. This work was supported by NIH GM-42136 (M.J.C.), a fellowship from the David and Lucille Packard Foundation (M.J.C.), and a NSF National Young Investigator (M.J.C.). C.J.G. and M.J.S. were supported by National Research Service Awards and G.P. was supported by a Brown-Coxe Fellowship.

References

Brewer, C. B., and Roth, M. G. (1991). A single amino acid change in the cytoplasmic domain alters the polarized delivery of influenza virus hemagglutinin. *J. Cell Biol.* **114,** 413–421.

Brown, D. A., Crise, B., and Rose, J. K. (1989). Mechanism of membrane anchoring affects polarized expression of two proteins in MDCK cells. *Science* **245,** 1499–1501.

Caceres, A., Banker, G. A., and Binder, L. (1986). Immunocytochemical localization of tubulin and microtubule-associated protein 2 during the development of hippocampal neurons in culture. *J. Neursci.* **6,** 714–722.

Caplan, M. J. (1990). Biosynthesis and sorting of the sodium, potassium-ATPase. *In* "Regulation of Potassium Transport Across Biological Membranes" (L. Reuss, J. M. Russell, and G. Szabo, eds.), pp. 77–101. Univ. of Texas Press, Austin.

Caplan, M. J., and Matlin, K. S. (1989). Sorting of membrane and secretory proteins in polarized epithelial cells. *In* "Functional Epithelial Cells in Culture (K. S. Matlin and J. D. Valentich, eds.), pp. 71–127. Alan R. Liss, New York.

Collawn, J. F., Stangel, M., Kuhn, L. A., Esekogwu, V., Jing, S., Trowbridge, I. S., and Tainer, J. A. (1990). Transferrin receptor internalization sequence YXRF implicates a tight turn as the structural recognition motif for endocytosis. *Cell* **63,** 1061–1072.

Cooley, L., Berg, C., and Spradling, A. (1988). Controlling P element insertional mutagenesis. *TIG* **4,** 254–258.

De Pont, J. J. H. H. M., Helmich-de Jong, M. L., Skrabanja, A. T. P., and van der Hijden, H. T. W. M. (1988). H,K-ATPase: Na,K-ATPase's stepsister. *In* "The Na,K Pump" (J. C. Skou, J. G. Norby, A. V. Maunsbach, and M. Esmann, eds.), pp. 585–602. Liss, New York.

Dotti, C. G., and Simons, K. (1990). Polarized sorting of viral glycoproteins to the axon and dendrites of hippocampal neurons in culture. *Cell* **62,** 63–72.

Dotti, C. G., Parton, R. G., and Simons, K. (1991). Polarized sorting of glypiated proteins in hippocampal neurons. *Nature (London)* **349,** 158–161.

Girones, N., Alvarez, E., Seth, A., Lin, I. M., Latour, D. A., and Davis, R. J. (1991). Mutational analysis of the cytoplasmic tail of the human transferrin receptor: Identification of a sub-domain that is required for rapid endocytosis. *J. Biol. Chem.* **266,** 19006–19012.

Goldstein, J. L., Brown, M. S., Anderson, R. G. W., Russell, D. W., and Schneider, W. J. (1985). Receptor-mediated endocytosis: Concepts emerging from the LDL receptor system. *Annu. Rev. Cell Biol.* **1,** 1–39.

Goslin, K., Schreyer, D. J., Skene, H. P., and Banker, G. (1990). Changes in the distribution of GAP-43 during the development of neuronal polarity. *J. Neurosci.* **10,** 588–602.

Gottardi, C. J., and Caplan, M. J. (1993a). An ion transporting ATPase encodes multiple apical localization signals. *J. Cell Biol.* **121,** 283–293.

Gottardi, C. J., and Caplan, M. J. (1993b). Molecular requirements for the cell surface expression of multisubunit ion-transporting ATPases: Identification of protein domains that participate in Na,K-ATPase and H,K-ATPase subunit assembly. *J. Biol. Chem.* **268,** 14342–14347.

Gottlieb, T. A., Gonzalez, A., Rizzolo, L., Rindler, M. J., Adesnik, M., and Sabatini, D. D. (1986). Sorting and endocytosis of viral glycoproteins in transfected polarized epithelial cells. *J. Cell Biol.* **102,** 1242–1255.

Guastella, J., Nelson, N., Nelson, H., Czyzyk, L., Kenyan, S., Miedel, M., Davidson, N., Lester, H., and Kanner, B. (1990). Cloning and expression of a rat brain GABA transporter. *Science* **249,** 1303–1306.

Hammerton, R. W., Krzeminski, K. A., Mays, R. W., Ryan, T. A., Wollner, D. A., and Nelson, W. J. (1991). Mechanism for regulating cell surface distribution of Na,K-ATPase in polarized epithelial cells. *Science* **254,** 847–850.

Iversen, L. L. (1971). Role of transmitter uptake mechanisms in synaptic neurotransmission. *Br. J. Pharmacol.* **41,** 571–591.

Jones, L. V., Compans, R. W., Davis, A. R., Bos, T. J., and Nayak, D. P. (1985). Surface expression of influenza virus neuraminidase, an amino-terminally anchored viral membrane glycoprotein in polarized epithelial cells. *Mol. Cell. Biol.* **5,** 2181–2189.

Jorgensen, P. L. (1982). Mechanism of the Na,K pump: Protein structure and conformations of the pure Na,K-ATPase. *Biochim. Biophys. Acta* **694,** 27–68.

Kordely, E., and Bennett, V. (1991). Distinct ankyrin isoforms at neuron cell bodies and nodes of Ranvier resolved using erythrocyte ankyrin-deficient mice. *J. Cell Biol.* **114,** 1243–1259.

Lentz, T. (1971). "Cell Fine Structure." Saunders, Philadelphia.

Lisanti, M. P., Le Bivic, A., Saltiel, A. R., and Rodriguez-Boulan, E. J. (1990). Preferred apical distribution of glycosyl-phosphatidylinositol (GPI) anchored proteins: A highly conserved feature of the polarized epithelial cell phenotype. *J. Membr. Biol.* **113,** 155–167.

Mandel, L. J., and Balaban, R. S. (1981). Stoichiometry and coupling of active transport to oxidative metabolism in epithelial tissues. *Am. J. Physiol.* **240,** F357–F371.

McDonough, A. A., Geering, K., and Farley, R. A. (1990). The sodium pump needs its beta subunit. *FASEB J.* **4,** 1589–1605.

McGrail, K. M., Phillips, J. M., and Sweadner, K. J. (1991). Immunofluorescent localization of three Na,K-ATPase isozymes in the rat central nervous system: Both neurons and glia can express more than one Na,K-ATPase. *J. Neurosci.* **11,** 381–391.

Mercier, F., Reggio, H., Devilliers, G., Bataille, D., and Mangeat, P. (1989). Membrane-cytoskeleton dynamics in rat parietal cells: Mobilization of actin and spectrin upon stimulation of gastric acid secretion. *J. Cell Biol.* **108,** 441–453.

Nelson, W. J. (1992). Regulation of cell surface polarity from bacteria to mammals. *Science* **258,** 948–955.

Nelson, W. J., and Hammerton, R. W. (1989). A membrane–cytoskeletal complex containing Na,K-ATPase, ankyrin and fodrin in Madin–Darby canine kidney cells: Implications for the biogenesis of epithelial cell polarity. *J. Cell Biol.* **108,** 893–902.

Nelson, W. J., and Veshnock, P. J. (1987). Ankyrin binding to Na,K-ATPase and implications for the organization of membrane domains in polarized cells. *Nature (London)* **328**, 533–536.

Okamoto, C. T., Karpilow, J. M., Smolka, A., and Forte, J. G. (1990). Isolation and characterization of gastric microsomal glycoproteins: Evidence for a glycosylated beta subunit of the H,K-ATPase. *Biochim. Biophys. Acta* **1037**, 360–372.

Okusa, M. D., Gottardi, C. J., Rajendran, V., Binder, H., and Caplan, M. J. (1994). Expression of a protein in kidney and colon that is related to the gastric H,K-ATPase. *Cell. Physiol. Biochem.* in press.

Pfaller, W., Gstraunthalerand, G., and Loidl, P. (1990). Morphology of the differentiation and maturation of LLC-PK$_1$ epithelia. *J. Cell. Physiol.* **142**, 247–254.

Pietrini, G., Matteoli, M., Banker, G., and Caplan, M. J. (1992). Isoforms of the Na,K-ATPase are present in both axons and dendrites of hippocampal neurons in culture. *Proc. Natl. Acad. Sci. U.S.A.* **89**, 8414–8418.

Primakoff, P., and Myles, D. G. (1983). A map of the guinea pig sperm surface constructed with monoclonal antibodies. *Dev. Biol.* **98**, 417–428.

Radian, R., and Kanner, B. (1983). Stoichiometry of sodium and chloride-coupled γ-aminobutyric acid transport by synaptic plasma membrane vesicles isolated from rat brain. *Biochemistry* **22**, 1236–1241.

Reuben, M. A., Lasater, L. S., and Sachs, G. (1990). Characterization of a beta subunit of the gastric H,K-ATPase. *Proc. Natl. Acad. Sci.* **87**, 6767–6771.

Rodman, J. S., Kerjaschki, D., Merisko, E., and Farquhar, M. G. (1984). Presence of an extensive clathrin coat on the apical plasmalemma of the rat kidney proximal tubule cell. *J. Cell Biol.* **98**, 1630–1636.

Rodriguez-Boulan, E. J., and Nelson, W. J. (1989). Morphogenesis of the polarized epithelial cell phenotype. *Science* **245**, 718–725.

Rodriguez-Boulan, E. J., and Pendergast, M. (1980). Polarized distribution of viral envelope glycoproteins in the plasma membrane of infected epithelial cells. *Cell* **20**, 45–54.

Rodriguez-Boulan, E. J., and Sabatini, D. D. (1978). Asymmetric budding of viruses in epithelial monolayers: A model system for study of epithelial polarity. *Proc. Natl. Acad. Sci. U.S.A.* **75**, 5071–5075.

Rose, J. K., and Doms, R. W. (1988). Regulation of protein export from the endoplasmic reticulum. *Annu. Rev. Cell Biol.* **4**, 257–288.

Roth, M. G., Compans, R. W., Giusti, L., Davis, A. R., Nayak, D. P., Gething, M. J., and Sambrook, J. S. (1983) Influenza virus hemeagglutinin expression is polarized in cells infected with recombinant SV40 viruses carrying cloned hemeagglutinin DNA. *Cell* **33**, 435–443.

Schultz, S. G. (1986). Cellular models of epithelial ion transport. *In* "Physiology of Membrane Disorders" (T. E. Andreoli, J. F. Hoffman, D. D. Fanestil, and S. G. Schultz, eds.), pp. 519–534. Plenum, New York.

Shull, G. E. (1990). cDNA cloning of the β-subunit of the rat gastric H,K-ATPase. *J. Biol. Chem.* **265**, 12123–12126.

Shull, G. E., and Lingrel, J. (1986). Molecular cloning of the rat stomach H,K-ATPase. *J. Biol. Chem.* **261**, 16788–16791.

Shull, G. E., Greeb, J., and Lingrel, J. B. (1986). Molecular cloning of three distinct forms of the Na,K-ATPase α-subunit from rat brain. *Biochemistry* **25**, 8125–8132.

Simons, K., and Fuller, S. D. (1985). Cell surface polarity in epithelia. *Annu. Rev. Cell Biol.* **1**, 295–340.

Simons, K., and Wandinger-Ness, A. (1990). Polarized sorting in epithelia. *Cell* **62**, 207–210.

Smith, P. R., Bradford, A. L., Joe, E. H., Anjelides, K. J., Benos, D. J., and Saccomani, G. (1993). Gastric parietal cell H,K-ATPase microsomes are associated with isoforms of ankyrin and spectrin. *Am. J. Physiol.* **264**, C63–C70.

Smolka, A., and Weinstein, W. A. (1986). Immunoassay of pig and human gastric proton pump. *Gastroenterology* **90,** 532–539.

Soroka, C. J., Chew, C. S., Modlin, I. M., D. M. Hanzel, Smolka, A., and Goldenring, J. (1993). Characterization of membrane and cytoskeletal components in cultured parietal cells using immunofluorescence and confocal microscopy. *Eur. J. Cell Biol.* **60,** 76–87.

Stephens, E. B., Compans, R. W., Earl, P., and Moss, B. (1986). Surface expression of viral glycoproteins is polarized in epithelial cells infected with recombinant vaccinia viral vectors. *EMBO J.* **5,** 237–245.

Sweadner, K. J., and Goldin, S. M. (1980). Active transport of sodium and potassium. *N. Engl. J. Med.* **302,** 777–783.

Yamauchi, A., Kwon, H. M., Uchida, S., Preston, A., and Handler, J. S. (1991). Myo-inositol and betaine transporters regulated by tonicity are basolateral in MDCK cells. *Am. J. Physiol.* **261,** F197–F202.

Yamauchi, A., Uchida, S., Kwon, H. M., Preston, A. S., Robey, R. B., Garcia-Perez, A., Burg, M. B., and Handler, J. S. (1992). Cloning of a Na and Cl-dependent betaine transporter that is regulated by hypertonicity. *J. Biol. Chem.* **267,** 649–652.

Index

ISBN 0-12-153341-7

90018

9 780121 533410